T0281219

Foundations in
Applied Nuclear
Engineering Analysis

2nd Edition

Foundations in
Applied Nuclear
Engineering Analysis

2nd Edition

Glenn E Sjoden
Georgia Institute of Technology, USA

 **World Scientific**

NEW JERSEY • LONDON • SINGAPORE • BEIJING • SHANGHAI • HONG KONG • TAIPEI • CHENNAI

Published by

World Scientific Publishing Co. Pte. Ltd.

5 Toh Tuck Link, Singapore 596224

USA office: 27 Warren Street, Suite 401-402, Hackensack, NJ 07601

UK office: 57 Shelton Street, Covent Garden, London WC2H 9HE

Library of Congress Cataloging-in-Publication Data
Sjoden, Glenn E., 1962–
    Foundations in applied nuclear engineering analysis / Glenn E Sjoden (Georgia Institute of Technology, USA). -- 2nd edition.
        pages cm
    Includes bibliographical references and index.
    ISBN 978-9814630924 (hardback : alk. paper) -- ISBN 978-9814630931 (pbk. : alk. paper) -- ISBN 978-9814630948 (ebook)
    1. Nuclear engineering--Mathematics--Textbooks. I. Title.
    TK9145.S535 2014
    621.48--dc23

                                    2014030553

British Library Cataloguing-in-Publication Data
A catalogue record for this book is available from the British Library.

In-house Editor: Amanda Yun

Typeset by Stallion Press
Email: enquiries@stallionpress.com

Printed in Singapore

To My wife Patricia and Children:
Alli, Daniel, and Nicole, and
to friends and family who were supportive,
and to God for the strength to complete this task

Preface to Second Edition

This text was developed as part of a *very* applied course in mathematical physics methods for nuclear engineers. The course in *Nuclear Engineering Analysis* that follows this text began at the University of Florida, and was later re-written while I served as a Professor of Nuclear Engineering at the Georgia Institute of Technology. The book covers a fast-paced one semester course to address applied concepts in mathematics, engineering analysis, and computational problem solving needed in subjects such as radiation interactions, heat transfer, reactor physics, radiation transport, numerical modeling, etc., for success in a nuclear engineering/medical physics curriculum. While certain topics are covered tangentially, others are covered in depth to target the appropriate amalgam of topics for success in navigating nuclear-related disciplines. It is assumed that students have familiarity with undergraduate engineering mathematics and physics, and are ready to apply those skills here; applications and problem sets in the appendix are directed toward problems in nuclear science. Software examples and programming are emphasized throughout the course, since computational capabilities are essential for new engineers. I would like to acknowledge assistance in typing some of the equations and content by Michael Wayson, and some contributions of content by Eric LaVigne. I also thank my students over the last 10 years for helping to edit content and provide feedback. Finally, I thank God and my family for the support they provided me in assembling the original text, and the second edition, with expanded topics and content. My goal has always been to better educate our future engineers for a better tomorrow.

G. E. Sjoden
Georgia Tech
Atlanta, Georgia

About the Author

Dr. Glenn Sjoden wrote the second edition of this text while on the faculty at the Georgia Institute of Technology. His career nuclear engineering experience spans a broad range of applications, having served in numerous capacities: technical director, nuclear research officer, professor, lead design engineer, and licensed engineering consultant.

During his military career, Dr. Sjoden served as an Air Force nuclear research officer from 1984 to 2004, which included three separate assignments working on treaty monitoring missions with the Air Force Technical Applications Center (AFTAC), and as a United States Air Force Academy faculty member. In 2004, Dr. Sjoden retired as a lieutenant colonel after 20 years of active service.

From 2004 to 2010, he served as a faculty member of the University of Florida (UF) in Nuclear and Radiological Engineering, as the UF Florida Power & Light Endowed Term Professor for Nuclear Power Research. From 2010 to 2014, Dr. Sjoden served as a tenured professor of Nuclear and Radiological Engineering in the George W. Woodruff School at the Georgia Institute of Technology in Atlanta, and as a Joint Faculty with Oak Ridge National Laboratory, and also as the Director of Georgia Tech's Radiological Science and Engineering Laboratory. During his 10 years in academia, Dr. Sjoden also served as an international consultant, and is the principal developer of the PENTRAN 3-D parallel deterministic radiation transport code. In July 2014, Dr. Sjoden became the Chief Scientist, AFTAC, at Patrick Air Force Base, Florida.

Contents

Chapter 1

Nuclear Engineering Analysis

Nuclear engineering analysis is a very broad subject title. Nuclear engineering and medical physics are disciplines that are ultimately very applied; while the "pure sciences" are often afforded the luxury of only considering hypothetical problems, nuclear engineers must respond to delivering real engineering systems to benefit humanity, e.g. providing nuclear power, advanced imaging devices, curing cancer, and the like. Therefore, nuclear engineers must design systems and devices that are based on nuclear interactions using subatomic particles that are not detectable by the ordinary human senses.

A fundamental and practical understanding of mathematics and physics must synergistically be employed to be effective as a nuclear engineer or medical physicist. The successful nuclear engineer must be adaptable, and able to create appropriate mathematical models, so that those models may be used to answer questions and solve real engineering problems. How does one achieve this? The answer is education, training, hard work, and practice.

In most cases, the inherent complexity of the systems involved in nuclear related disciplines require computation. Models are often implemented as computer programs; almost every "real" problem in nuclear engineering requires the use of a computer in some manner, and programming skills among any variety of languages and tools is essential. It is envisioned that users of this text will be assigned problems that involve computation, particularly as progress through the material culminates in solving partial differential equations.

Origins of Nuclear Engineering

Nuclear Engineering began its historical roots in 450 B.C. Greece, where Democritus argued that substances were ultimately composed of small, indivisible particles that he labeled "atoms." In 1869, Russian chemist Mendeleev organized these elements of particular atoms into a table that grouped them by their physical properties and characteristics called "the periodic table of the elements". Much work was advanced over time so that in the early 20th Century, and nuclear fission was discovered by Hahn & Strassman in 1939, where Barium and Krypton resulted from bombarding neutrons into Uranium, a conclusion reached with counsel from scientist Lise Meitner.

An American named William Arnold, visiting Otto Frisch to view a confirmatory experiment, recalled that "binary fission" was a biology term describing when one Bacterium divides into two... hence "nuclear fission" came from liberating energy by splitting the atom, and with it, nuclear engineering was born. (P.260-266 Rhodes, "Making of the A-bomb").

Nuclear energy always invokes "mysterious wonder" from many, since its initial release was a result of the then very secret "Manhattan Project" and atom bomb development during World War II. For many years and into present day, nuclear energy and engineering principles maintain a "mystery quality" due to an association with nuclear weapons. Granted, the awesome amount of energy release possible from fission and subsequent weapons development during the Cold War to present day furthered this image; an example of weapons developed is given in the following figure.

"Grable" 280 mm howitzer nuclear cannon test, Nevada Test Site (US Govt image)

Modern applications of nuclear engineering are focused primarily on power and medical applications, as noted. For nuclear power generation, most of these involve reactor design. While a nuclear weapon is a device designed to release nuclear energy using an uncontrolled fission chain reaction, a nuclear reactor is a device in which nuclear energy from fission is released in a controlled manner using fission neutrons as the fission chain carrier.

The principal subatomic particles are the *proton* (charge of $+1$, 1.007277 amu, where 1 amu = 1.66054 E-27 kg = 1/12 of a carbon atom), the *neutron* (neutral charge, 1.008665 amu), and the electron (charge of -1, 5.4858E-4 amu). The operation of a reactor relies on the fate of subatomic particles (neutrons) released in the reactor as a consequence of fission. Fission is brought about by neutron irradiation of fissile materials; when a neutron is absorbed by a fissile nucleus, this results in an unstable excited nucleus that splits into two smaller, more stable pieces while liberating typically 2 to 3 new neutrons in each fission, which then go on to further sustain the chain reaction or leak out. Not counting the neutrons, the two large halves or "fission fragments" are often radioactive. Therefore, neutrons interact with nuclei in the reactor system and serve as a "fission chain carrier," governing how the fission reactions continue. Therefore, understanding "neutronics," or neutron balance in a reactor, is a fundamental principle of "Reactor Physics," and we will expand upon this late in the text, where we will consider the neutron transport and diffusion problem for some fundamental scenarios.

Nuclear Applications

Nuclear fission reactors can produce power for electricity generation, $_1^1H$ generation for fuel cell based transportation, nuclear driven propulsion in ships and submarines, and even space travel, including nuclear rocket engines, compact nuclear power systems, etc. Medical devices use nuclear reactions include linear accelerators of various types where radioactive particles are used to kill cancer cells in humans, and generate isotopes used in diagnosis, imaging, and cancer treatment.

Work to support applications in nuclear energy and medical applications requires a wide range of applied mathematics, which was the motivation for writing this text... students need to be exposed to a variety of applications to reinforce their ability to be problem solvers! To be good at problem solving, for nuclear applications, this usually involves some type of computing and computer programming; so we briefly discuss this in the next section.

Computer Programming

Computer programming plays several important roles in nuclear engineering analysis: modeling problems, exploring new ideas, automating well-known techniques, and creating new tools for other engineers. A programming language is a tool for creating computer programs. As engineers, we should be aware of each tool's strengths so that we can choose the right tool for each job. In this book, a few programming languages are noted: *Mathematica*, *TK-Solver*, and FORTRAN or C. Programming is an essential tool for nuclear engineers; mastering this in a variety of forms yields new and profound understanding of the physics.

Mathematica is an advanced commercial programming language with built-in support for common tasks in science, engineering, and mathematics. *Mathematica* is particularly well-suited to trying out new ideas or profiling your homework. This tool is expensive, but worth the price (especially with a student version discount).

TK-Solver is a useful tool for modeling engineering problems and solving systems of equations. One advantage of TK-Solver is that it is often possible to model new systems without doing any new programming. Variables can be interchangeably mixed at will as input or output, and therefore TK-Solver models can be readily used as "design optimization" tools.

FORTRAN or **C** languages are often used for programs that need to be fast, such as for neutron transport simulators. You may need to know FORTRAN if you are trying to improve processing data from a "legacy" nuclear application, because FORTRAN has always been popular among nuclear engineers; many codes that have been "nuclear certified" in FORTRAN may never be re-written in other languages, since certification in a new language is cost-prohibitive. Therefore, it is envisioned that FORTRAN will be alive and well in the foreseeable future simply due to its widespread (and certified) use in the nuclear industry. Message Passing Interface (MPI) libraries have enabled FORTRAN and C codes to be readily parallelized on multiprocessor MIMD (multiple instruction, multiple data) computers.

Later, if you find yourself working on a programming project which will last more than a month, it would be worthwhile to spend a week mastering a programming language that is well-suited to a particular task. Choosing the right tool can speed up the work, and often also results in a higher quality result. Flexibility in programming is an important key for success, and some alternative programming tools are OCaml, Python, and a few others.

OCaml, and its Microsoft equivalent **F#**, are similar to Fortran, but with many improvements. Arrays are easier to use, and there is support for additional data structures, which are helpful when optimizing a program for large data sets. Also, OCaml programs tend to be shorter than similar programs written in Fortran.

Python is well-suited for creating websites [Django] or graphical user interfaces and using programs that were written in other languages. Python is very easy to learn, and Python programs can often be understood by engineers who have never used Python. **Numpy** is numerical python, a powerful library package imported in python that expands the use of python as a numerical solver.

Erlang makes it much easier to create programs that are distributed across many computers. Erlang is also capable of using programs that were written in other languages, so a high-performance computer program could use OCaml for fast calculations while Erlang takes care of coordinating tasks between computers.

R was designed specifically for statistical analysis and is also a good choice for plotting. Engineers who expect to dedicate many years of their careers to computer programming may also learn **Scheme**, **Smalltalk**, and **Haskell** to broaden their understanding of what is possible in computing.

Any of the above languages have their advantages and disadvantages. In any event, nuclear engineers must be adaptable programmers to be effective, even if to successfully navigate their educational necessities. Most problems of real interest require computational support. It is the responsibility of each nuclear engineer to be able to successfully develop and execute programs. Supercomputing with parallel processing is now a main-stream tool, and in the next section, this is briefly discussed.

Parallel Processing/Supercomputing

The idea of multiprocessing and the evolution of parallel computing began in the late 1950s, and continues today. Traditional "Serial" programming ("von Neumann" programming) is the execution of sequential tasks, one by one, in an ordered relationship of instructions. Serial processors have leaped ahead according to *Moore's Law*; Intel co-founder Gordon E. Moore established *Moore's Law* in his 1965 paper, "Cramming more components onto integrated circuits," *Electronics Magazine*, noting that the number of transistors on integrated circuits doubles approximately every two years. No matter how many flops or "floating point operations per second," can be achieved by a single processor, to get even more processing power massed together to solve large problems, one needs to *parallelize*. It is a lot like cutting grass on a large acreage. It is not cost effective to build a "large custom lawn mower" to cut the grass; rather, it is efficient to obtain several efficient, mass produced (low cost) lawn mowers, and have several people cutting the grass at the same time. This is essentially how parallel computing is accomplished; to have effective parallel computing, one has to have a parallel program that can execute on many processors at the same time.

Therefore, "parallel" programming is the adaptation of an algorithm to distribute tasks that do not specifically rely on a prescribed order, so that these tasks may be executed simultaneously on multiple processors, with the overall result applied at the end.

Minimal synchronization between the processors is desired, since *any* synchronization amounts to "parallel overhead." Anything using the word "overhead" implies inefficiency and cost, and is therefore something to avoid, although everything in computing (and in life in general) has some amount of overhead associated with it that we must contend with. In any case, if we can avoid un-necessary synchronizations, we can achieve a high parallel efficiency. Due to the ten-fold increase in computational performance during each five year period as computing became main stream, multiprocessing has made great progress.

The reasons for attempting to solve any numerically intensive problem with multiprocessing are simple: to reduce execution time, obtain higher accuracy, and/or solve problems that are larger than can be solved using a traditional, single processor von Neumann architecture (Freeman and Phillips, 1992). It is useful to define some common parallel processing terms; they are only briefly mentioned here to familiarize the reader with key terms.

Machine Classes and Memory Types

There are four classes of machines, as introduced by Flynn (1972):

- SISD — Single Instruction/Single Data Stream (a traditional "von Neumann" machine)
- SIMD — Single Instruction/Multiple Data Stream (lock-step arrays, vector machines)
- MISD — Multiple Instruction/Single Data Stream (all tasks contribute to one data set)
- MIMD — Multiple Instruction/Multiple Data Stream (multiple independent tasking)

Shared memory MIMD systems are constructed so that each processor has global memory access and are typically limited to tens of processors due to the large number of physical connections to the memory map.

Distributed memory MIMD parallel computers maintain completely independent memories, where processors exchange information by message passing over a high speed network; each processor independently executes code and can perform independent input/output (I/O) if allowed for in the parallel algorithm. Distributed memory processors, viewed as "nodes," are typically connected together using a variety of topologies, and can range in number into the thousands.

Speedup and Efficiency

Parallel performance models are necessary for analyzing and quantifying parallel speedup and efficiency. Parallel *speedup* (S_P) measures the overall reduction in computing time to solve a problem. It is defined as the wall-clock time on a serial (single) processor divided by the wall-clock time on P processors:

$$S_P = T_S / T_P$$

Parallel *efficiency* (E_P) measures the economic advantage of the parallelization by comparing the speedup factor to the allocated number of processors (Freeman and Phillips, 1992):

$$E_P = S_P / P$$

It is assumed here that the parallel algorithm *overhead*, the extra executable code and storage required to expedite parallel execution, is negligible during single processor execution. This is a typical convention often adopted in measuring parallel speedup (Werner, 1981).

Amdahl's Law

Also, an upper bound on anticipated parallel speedup can be determined by applying the *Amdahl's Law*, which states that given the fraction of a code that is parallelizeable: $0 < f_P < 1$, the maximum observed speedup for P processors with parallel communication time (T_C) is equal to:

$$S_P = \frac{1}{(1 - f_P) + f_P / P + T_C / T_S}$$

where in the limit of an infinite number of processes (assuming zero communication time):

$$\lim_{p \to \infty} S_P \to \frac{1}{(1 - f_P)}$$

Therefore, from the above equation, if the parallelizeable portion of a code is $f_P = 0.80$, the maximum *theoretically* observed speedup is 5.0, regardless of the number of additional processors added to the problem. In reality, due to increasing parallel instruction and communication overhead with the addition of more and more processors, there will be a point (depending on f_P, system architecture, and problem size) where adding more processors leads to extremely low efficiencies. This may be irrelevant if the code is scalable in memory, where, regardless of speed, the problem requires some number of processors to be solved *at all*. A plot of Amdahl's law (assuming $Tc = 0$) depicting maximum theoretical speedup based on parallel fraction and associated efficiency as a function of processors is provided for illustration below.

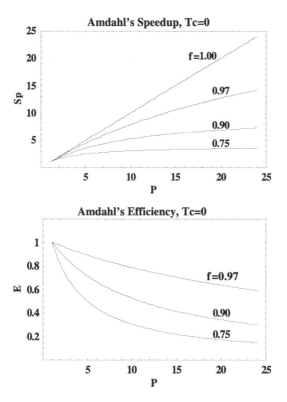

Amdahl's Law, yielding maximum theoretical speedup (Top) and parallel efficiency (Bottom), depicted for various parallel code fractions (f) and number of processors (P).

Load Balancing and Granularity

Other important terms include *load balancing* and *granularity*. Load balancing involves distributing work to processors evenly to maximize parallel efficiency. Algorithm granularity is a qualitative term that refers to the number of process operations that can be executed by each processor before a synchronization (or communication) of the processors must be implemented. These synchronizations can be viewed as serial barriers that limit parallel performance. Conventional definitions of grain size are:

- fine grain — unit numbers of operations before synchronization
- medium grain — tens of operations before synchronization
- coarse grain — hundreds (or more) of operations before synchronization

Computation/Communication Ratio and Scalability

The computation/communication ratio varies from machine to machine; this is the ratio of the CPU instruction speed in flops (floating point operations per second speed, often stated in Mflops, or millions of flops) attainable on the system to the relative speed of data transfer between processors. The speed of data transfer is related to communication latency, or the time required to send a zero byte-length message, and the communication bandwidth, in megabytes per second. The computation/communication ratio is often more of an issue on distributed memory, message passing machines, as network communication data rates are typically orders of magnitude slower than typical Mflop rates (Gropp, et al, 1994).

Dedicated vs. Non-dedicated computing is important when evaluating parallel performance. Parallel machine availability is a practical performance issue. If processors are available to users in a dedicated mode, then during parallel execution, a single user has complete control of the processors, and parallel performance can be accurately determined. Alternatively, in a non-dedicated (interactive) mode, processors are simultaneously available to many users; absolute parallel performance may be difficult to verify in this case.

With regard to a working definition of parallel scalability, scalable algorithms maximize computation to communication ratio, minimize serial operations, maximize algorithm and data parallelism, and maximize efficiency for the architecture (shared versus distributed memory) (Gerner, 1995). While there is a great deal more that can be mentioned about multiprocessing fundamentals and terminology, more complete discussions can be found elsewhere (see Freeman and Phillips, 1992, and Gropp, et al, 1994, and Chandy and Misra, 1988).

MPI (Message Passing Interface) Library

The Message Passing Interface library was initially established in the early 1990s with the "MPI Vendor Forum" and was formalized in 1995 as a library for creating algorithms to operate on distributed memory machines using standard languages (FORTRAN, C, other language variants supported). MPI enabled a standardization of parallel computing tasks. Modern "clusters" for multi-processing are now widespread in science, industry, and academia, and most are running some form of MPI. The following table depicts the minimum essential commands in which to call an MPI program.

Function	Purpose
MPI_INIT	Initiate MPI
MPI_COMM_SIZE	Find out how many processors are present
MPI_COMM_RANK	Determine the rank of the calling process
MPI_SEND	Send a message
MPI_RECV	Receive a message
MPI_FINALIZE	Terminate MPI

These commands can be integrated into a standard serial program that is altered with "parallel logic" so that all processors will run the code, but each core executing the code can perform different tasks that ultimately contribute to the calculation. Note that "Send" and "Receive" are the most basic communication modes—there are numerous other modes, and a good text on parallel programming, as well as numerous web sites with tutorials, should be reviewed to see additional examples and applications involved in writing parallel code. An example FORTRAN code "hello-bcast" is useful for illustration to demonstrate the "MPI_BCAST" parallel command is presented below. "MPI_BCAST" is the broadcast" command, which enables one processor among those participating to transmit a message to all other processors.

The "mpif.h" in the INCLUDE statement is what links in the libraries for MPI. The MPI_INIT is the first command to execute after the variable declarations. Recall that all processors are captured by the MPI_COMM_WORLD communicator. Here, MPI_COMM_WORLD represents all processors, and if there were 4 processors, they would be numbered 0, 1, 2, and 3, since all processor numbers begin with 0.

```
!****************************************************
program hello-bcast

include 'mpif.h'

integer n, myid, numprocs, i, rc

call MPI_INIT( ierr )
call MPI_COMM_RANK( MPI_COMM_WORLD, myid, ierr )
call MPI_COMM_SIZE( MPI_COMM_WORLD, numprocs, ierr )

WRITE(*,*) 'Process ', myid, ' of ', numprocs, ' is alive'

IF(myid.EQ.0)THEN
  n=100
ELSE
  n=0
END IF

WRITE(*,*) 'before: I am ',myid,' and n=',n
call MPI_BCAST(n,1,MPI_INTEGER,0,MPI_COMM_WORLD,ierr)
WRITE(*,*) 'after: I am ',myid,' and n=',n

call MPI_FINALIZE(ierr)
STOP

END
!****************************************************
```

Subgroups of processors captured by "custom" communicators can also be defined. Since the same code is executed by all processors, the logic to capture independent use of each processor is captured by the user in the style and architecture of the program.

Here, the process number is loaded into the variable "myid". Sometimes, it might prove useful to enable natural numbering from 1 to 4 reported under, for example, a new variable called "myidp" to declare a variable called "myidp=myid+1" after the MPI_COMM_RANK call. Again, it is up to the user to define this if desired. The next block contains an "IF-THEN" code block that alters the sequence for processor "0" but not the other processors. Then before the MPI_BCAST command is called, a message is written on each processor to report the "before" value of the integer n, and so is also done "after" the broadcast. Then, the code is complete. This simple code demonstrates how one can control the processes, all running the same code, with logic to perform parallel tasks. The output of this code is shown executed on 4 processors. Note that the values are in order reported by the processor participating as part of the MPI_COMM_WORLD communicator collective.

Process	3 of	4 is alive		
Process	1 of	4 is alive		
Process	0 of	4 is alive		
Process	2 of	4 is alive		
before: I am	1 and n=	0		
before: I am	3 and n=	0		
before: I am	2 and n=	0		
before: I am	0 and n=	100		
after: I am	0 and n=	100		
after: I am	1 and n=	100		
after: I am	2 and n=	100		
after: I am	3 and n=	100		

Radiation

What separates nuclear engineers from all other engineering disciplines is that nuclear engineers have the added requirement of uniquely understanding radiation, radiation interaction, and radioactive materials.

The word *Radiation* often refers to "ionizing" radiation, ionizing a gas through which the radiation passes. Other radiation, e.g. radio waves, is "non-ionizing". *Ionization* is the process in which a neutral atom or molecule is given a net electrical charge. The amount of energy to remove the least tightly bound e^- from an atom is the first ionization potential. In the early 1900's, many changes were occurring in Physics that would have a profound influence on 20th Century theory. In particular, discussions surrounding quantum theory and the electromagnetic spectrum were compelling. Discoveries included:

EM Waves, Electrons, X-rays, Natural Radioactivity, α (He nuclei), β (+,− electrons), and γ (photons) radiation, neutrons, and fission

Radiation Emission for Nuclear Engineers or Medical Physicists refers to a release of energy. Particles and waves that have energy actually have a "particle-wave" duality, as was shown by de Broglie..

A unit of measure of radioactivity is the Becquerel (Bq) and the Curie (Ci). One Bq = 1 disintegration per second (dis/s). One Ci = 3.7E10 dis/s, and the Curie is defined as that amount of radioactivity which has the same disintegration rate as 1 g of Radium-226. Variations on these units for activity of radioactive atoms are used interchangeably.

de Broglie Waves

Louis-Victor-Pierre-Raymond, 7th duc de Broglie, a French physicist best known for his research on quantum theory and for his discovery of the wave nature of electrons. de Broglie was awarded the 1929 Nobel Prize for Physics; he was born on Aug. 15, 1892, Dieppe, France, and died on March 19, 1987, in Paris. de Broglie established particle-wave duality of matter—in certain instances, matter acts as a particle; in other instances it acts as a wave. Both neutrons and gamma rays are neutral particles that can be also considered as waves based on de Broglie duality.

Classical Particle

Position (x, y, z) $\vec{r} = <x, y, z>$ position vector

Momentum $\vec{p} = m\vec{v}$

$E = h\nu = h\dfrac{c}{\lambda}$ where ν is the frequency and λ is the Wavelength

$$p = \left(\frac{hc}{\lambda}\right)\frac{1}{c} \Rightarrow \frac{h}{\lambda} \rightarrow \left[\lambda = \frac{h}{p}\right] \qquad \vec{p} = m\vec{v}$$

de Broglie proposed that any material body will have a wavelength associated with its momentum (motion)…Actually this has been proven for e^-'s diffracted through a slit, like waves. Given the de Broglie particle-wave duality, we can find the de Broglie wavelength of a 10 g pellet moving at 10 m/s (fired from a paintball gun).

$$\lambda = \frac{h}{mv} = \frac{6.63 \times 10^{-34} J \cdot s}{0.01 kg \cdot 10 m/s} \frac{\dfrac{kg \cdot m^2}{s^2}}{J}$$
$$= 6.63 \times 10^{-33} m$$

Since $1\mathring{A} = 10^{-10} m$, this is $6.63 \times 10^{-23} \mathring{A}$

Note that this is very small, in fact, it is too small to measure!

A more practical example of de Broglie's duality theory pertains to the radius of the nucleus. The radius of a nucleus is given by:

$$r_{nuc} = 1.2 \cdot 10^{-15} A^{1/3} \text{ meters, where } A \text{ is the mass number}$$

What minimum energy of photons might be used to probe an object $0.1 \times 10^{-9} m$ (0.1 nm) in size?

$$E = \frac{hc}{\lambda} = \frac{6.63 \times 10^{-34} J \cdot s}{0.1 \times 10^{-9}} \frac{3 \times 10^{8} m/s}{1.602 \times 10^{-16} J/keV} = 12.4 keV$$

the research of Planck, de Broglie, Einstein, and others led to the advent of Quantum Theory.

Quantum Theory

Quantum Theory—that energy can only be exchanged in "packets" or discrete amounts when information is exchanged from one state to another. With quantum theory comes the possibility that everything in physics has a finite probability. Some probabilities associated with events are very small; still, the likelihood is non-zero.

Nuclear Engineering deals with nuclear radiation processes that occur based on a *probability*. For example, neutrons bombarding nuclear fuel can cause fission, liberating "prompt fission neutrons"... those prompt neutrons are emitted within $10^{-17} s$ of a fission event... and there is a certain probability of neutron-nucleus interaction, called a *cross section*.

Cross Sections

Cross sections (often referred to as "nuclear data") can be stated in either *microscopic* or *macroscopic* form. The units of the cross section define *how* they are used; typically, nuclear interaction cross sections are denoted by the Greek symbol "sigma", be it either σ or Σ. Some texts adopt the convention that the lower case sigma be used as microscopic, and the upper case as macroscopic. However, it is the units defining the cross section that really define what the cross section means. Details on neutron cross sections, and microscopic and macroscopic cross sections are defined as follows:

- A *microscopic cross section* is the effective "target area" presented by a nucleus, and are typically given in barns (1 barn (b) $= 1.0 \times 10^{-24}$ cm^2).

- A *macroscopic cross section* accounts for the atomic density of nuclei in the substance, and is the effective "target area" presented by a nucleus to an incident neutron *per unit volume*.

- Cross sections averaged over a single energy range are "one group" cross sections; these are usually given as:

  o σ_s Scattering cross section, where the incident particle (neutron) scatters

  o σ_a Absorption cross section, where the incident particle is absorbed; absorption can be from a capture reaction or a fission reaction, and so the absorption cross section is the sum of σ_γ (capture) + σ_f (fission) cross sections.

  o σ_t Total cross section, where the incident particle (neutron) interacts through absorption or scattering; $\sigma_t = \sigma_a + \sigma_s$

  o Since neutrons or photons carry no charge, each can penetrate the nucleus easily without the nee to overcome Coulomb barrier effects; in that context, "neutral particle" interactions typically consider neutrons and photon effects.

- Cross sections averaged over a single energy range are "one group" cross sections; so-called "multi-group" cross sections are cross sections given as a function of different neutron energy widths or "bins," where cross sections are averaged over bins over a set of defined neutron energy values.

- If we use the convention that the lower case sigma σ be used as *microscopic*, and upper case Σ as *macroscopic*, then:

  o macroscopic case is equal to $\Sigma = N\sigma$, where N is the atom density. The macroscopic cross-section is typically given in units of cm^{-1}.

- Atom density comes from the density of atoms in the sample. This can be determined from:

$$N\left(\frac{atoms}{cm^3}\right) = \rho\left(\frac{g}{cm^3}\right)\frac{1}{A}\left(\frac{gmole}{g}\right)N_A\left(\frac{atom}{gmole}\right)$$

Where we note that N_A is Avogadro's number, $6.0226E+23$ atom/gmole. Also, "gmole" refers to "gram mole" of mass. Sometimes authors will use "kmole" which is a "kilogram mole". It is obviously very important to distinguish which units one is using. Most often, "gmole" is used.

Example

Elemental boron has a density of 2.34 g/cc. Determine the atom density, scatter, and total cross section if the atomic mass is 10.81 g/gmole and at room temperature, the microscopic scattering cross section is 4 b, and the absorption cross section is 669 b.

$$2.34\left(\frac{g}{cm^3}\right)\frac{1}{10.81}\left(\frac{gmole}{g}\right)6.02256E+23\left(\frac{atom}{gmole}\right)$$

$$= N_{Boron}\left(\frac{atom}{cm^3}\right) = 1.3037E+23$$

$$\Sigma_s\left(\frac{1}{cm^1}\right) = 1.3037E+23\left(\frac{atom}{cm^3}\right)(4)\left(\frac{b}{atom}\right)(1E-24)\left(\frac{cm^2}{b}\right)$$

$$\Sigma_s = 0.5215\left(\frac{1}{cm^1}\right)$$

Using this same approach the total macroscopic cross section of elemental boron is:

$$\Sigma_t = 87.74 \left(\frac{1}{cm^1} \right)$$

In any case, the amount of area presented by a nucleus for a neutron interaction can be directly correlated to a likelihood for interaction, in effect, a *probability*. Cross sections are used in some applications in this book, but the intent of this text is not to teach nuclear physics, or nuclear interactions, or reactor theory. The intent of material presented here is to cover the essential applied mathematics and physics topics needed for new students in nuclear engineering to gain experience and confidence in a limited yet focused set of analytic and numeric problem solving skills.

Atom and Weight Fractions

Often, cross sections are expressed in atom fraction or weight fraction. One must be adept at transferring between these two notations; if we consider how these are related, this is depicted below with atom fraction (af) (left) and weight fraction (wf) (right) as follows, demonstrated with the relative fractions of U-235 in natural uranium, with:

$$\frac{af_{25}M_{25}}{M_u} = wf_{25} \qquad \frac{wf_{25}M_u}{M_{25}} = af_{25}$$

$$M_u = 238.018 \qquad M_{25} = 235.044 \qquad af_{25} = 0.0072$$

$$\text{Then} \quad wf_{25} = (0.0072)(235.044)/(238.018) = 0.00711$$

$$\text{Also:} \quad M_u = af_{25}M_{25} + af_{28}M_{28}$$

The mass of the overall uranium comes from:

$$\frac{1}{M} = \sum_{all\ i} \frac{wf_i}{M_i}$$

Energy and Nuclear Reactions

Energy expressed in nuclear engineering is based on the electron volt, or eV—this is the energy acquired by a unit (electron) charge that is accelerated through an electric potential (electric field) of 1 volt. Nuclear reactions, including fission, are often expressed in MeV, or million electron volts −1 MeV is 1 million electron volts.

Putting this to immediate use, bombarding thermal neutrons on a uranium-235 nucleus will result in fission and release of ~200 MeV of energy, with 168 MeV allocated to fission fragments, 7 MeV allocated to gamma rays (6 to 8 gammas of energy of ~1 MeV), 5 MeV in neutrons (2.43 neutrons per fission), and the remaining energy split between fission products beta, gamma, and neutrino/anti-neutrino radiation.

When dealing with neutrons, protons, or fission products, in general we can achieve reasonable accuracy by assuming *classical* physics relationships hold; only when dealing with high energy electrons (small mass, with velocities close to the speed of light *c*) do we need to consider relativistic physics treatments (of course, gamma rays also travel at the speed of light). To demonstrate that classical physics treatment is reasonably accurate, we wish to consider the total energy of a particle, and thereby refer to Einstein's relativistic treatment, where the kinetic energy (*T*) of the particle is the total energy minus the rest energy (with rest mass m_0), as given below:

$$T = \frac{m_0 c^2}{\sqrt{1 - \dfrac{v^2}{c^2}}} - m_0 c^2$$

$$T = m_0 c^2 \left(\frac{1}{\sqrt{1 - \dfrac{v^2}{c^2}}} - 1 \right)$$

Then we have:

$$\left[\left(\frac{T}{m_0 c^2} + 1 \right) = \frac{1}{\sqrt{1 - \dfrac{v^2}{c^2}}} \right]$$

and re-arranging further:

$$\left(1 - \frac{v^2}{c^2} \right) = \frac{1}{\left(1 + \dfrac{T}{m_0 c^2} \right)^2} \quad \text{or} \quad \left(1 - \frac{1}{\left(1 + \dfrac{T}{m_0 c^2} \right)^2} \right) = \frac{v^2}{c^2}$$

which yields:

$$v = \sqrt{c^2 \left(1 - \frac{1}{\left(1 + \left(\dfrac{T}{m_0 c^2} \right) \right)^2} \right)}$$

Then assuming $T = 106.5\,MeV$ for the ^{87}Kr fission fragment, and

$$m_0 = 1.661x10^{-27}\,\frac{kg}{amu} \cdot 87\,amu = 1.4503\,x10^{-25}\,kg$$

$$v = \sqrt{\left(3x10^8\,\frac{m}{s}\right)^2 \left(1 - \frac{1}{\left(1 + \dfrac{106.5\,MeV \cdot 1.602\,x10^{-13}\,\dfrac{J}{MeV}}{1.4503\,x10^{-25}\,kg\left(3x10^8\,\dfrac{m}{s}\right)^2}\right)^2}\right)}$$

$$\boxed{v = 1.5324\,x\,10^7\,\frac{m}{s}}\ \ \text{by relativistic physics.}$$

$$T_{classical} = \frac{1}{2}m_0 v^2$$

$$106.5\,MeV \cdot 1.602\,x\,10^{-13}\,\frac{J}{MeV} = \frac{1}{2} \cdot 1.4503\,x\,10^{-25}\,kg \cdot v^2$$

$$\boxed{v = 1.5338\,x\,10^7\,\frac{m}{s}}\ \ \text{by classical physics.}$$

Note that these two predicted speeds are ~0.1% apart, so that classical physics affords reasonable accuracy for most applications with nucleons of sufficient mass in the energies spanning nuclear fission.

NUCLEAR APPLICATION: Binding Energy

When considering nuclear reactions, in all cases, the fundamental physics of the conservation of mass, energy, and momentum must always apply. The mass defect (Δ) is the difference between the mass of a nucleus and the sum of the masses of the nucleons of which the nucleus is composed. In combining, some of the mass is converted to energy to tightly bind the components. This mass defect is calculated as follows in terms of nuclear masses, where M_p is the proton mass, and M_n is the neutron mass, and $M_{nucleus}$ is the mass of the stable nucleus, Z is the atomic number (denoting charge), and A is the mass number:

$$\Delta = \left(Z M_p + (A-Z) M_n \right) - M_{nucleus}$$

Most measurement data is based on *atomic* mass data, so by adding and subtracting $Z M_e$ (where M_e is the electron mass):

$$\Delta = \left(Z M_p + (A-Z) M_n \right) - Z M_e - M_{nucleus}$$

Reformulating, we can cast the relationships in terms of the mass of a hydrogen atom M_H and the mass of the atom, M_{Atom}:

$$\Delta = \left(Z \left(\underbrace{M_p + M_e}_{M_H + BE_{e-}} \right) + (A-Z) M_n \right) - \left(\underbrace{M_{nucleus} + Z M_e}_{M_{Atom} + BE_{ze-}} \right)$$

$$\Downarrow \qquad\qquad\qquad\qquad \Downarrow$$

electron binding energy
in H atom
$\qquad\qquad$ electron binding energy
of atom with Z, A

So that we obtain:

$$\Delta = \left(Z M_H + (A-Z) M_n - M_{Atom} + \left(\underbrace{Z BE_{e-} - BE_{ze-}}_{Tend\ to\ cancel} \right) \right)$$

Then, neglecting the respective hydrogen atom electron binding energies and binding energies of Z electrons in the atom, these two terms tend to approximately cancel: $\left(ZBE_{e-} - BE_{ze-}\right)$, resulting in:

$$\Delta \equiv Z M_H + \left(A-Z\right)M_n - M_{Atom}$$

where $M_n = 1.00866544\ amu$, $M_H = 1.00782522\ amu$

This mass defect can be cast in terms of energy with the aid of the Einstein Equation $E = mc^2$, or more formally

$$E = mc^2 = \frac{m_0}{\sqrt{1 - \dfrac{v^2}{c^2}}} c^2$$

where m_0 is the rest mass of the particle and v is the velocity relative to c, the speed of light, $2.997E{+}08\ m/s$. From this, one atomic mass unit (amu) is equivalent to 931.494 MeV. If we calculate the binding energy per nucleon (neutrons and protons), we see the "curve of binding energy":

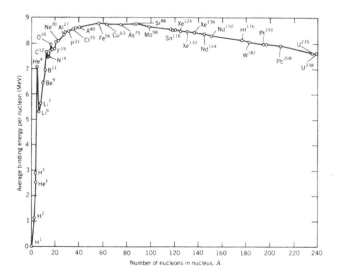

Binding energy per nucleon (public domain image)

This means that mid-level mass numbers of nucleons are more "tightly bound" since it would require more binding energy to supply to these nuclides in order to break them apart into their separate nucleons. As a result, one can see that carrying out a fission reaction, splitting the very heavy nuclides into approximately two halves of the original mass, such as U-235, or combining the light nuclides in a fusion reaction, such as with H-2 and/or H-3 (deuterium, d, and/or tritium, T) essentially doubling the mass, will result in more stable nuclides following each reaction.

NUCLEAR APPLICATION: Nuclear Reactions

In nuclear reactions, these are established by a convention detailing reaction particles:

$$A \ (B, C) \ D$$

Where:
- A – Target; Nucleus target, often considered at rest compared to incident energy (speed) of bombarding particle
- B – Projectile; Bombarding particle incident on target nucleus
- C – Product Projectile; Exiting particle, after formation of a compound nucleus
- D – Product Nucleus; Recoiling nucleus, post reaction product

The so called "Q-Value" gives us the relative change in energy for a particular reaction, and enables us to determine, based on the initial conditions, whether or not the reaction will occur freely with enough energy, *exothermically*, with a positive value, or will require an energy input, *endothermically*, and yield a negative Q value:

$$Q = \text{Reactants} - \text{Products}$$

This is best demonstrated by an example.

Example

Consider the fusion reaction where a deuterium nucleus collides with a tritium atom, neutrons are rendered from the reaction:

$$T(d,n)He_2^4$$

The reaction, written in long form, is:

$$d_1^2 + T_1^3 \rightarrow n_0^1 + He_2^4$$

Where it should be noted that the nucleon charges and mass numbers balance. Also, $M_d = 2.014102$ amu; $M_T = 3.016049$ amu; $M_{He} = 4.002604$ amu; $M_n = 1.008665$ amu.

The Change in the kinetic energies of the particles before and after the reaction is equal to the difference in the rest mass energies of the reaction components before and after the reaction:

$$Q = (M_d + M_T) - (M_n + M_{He})$$

$$Q = (5.030151) - 5.011269) = 0.018882 \text{ amu}$$

Recall the mass-energy equivalent, from the Einstein relation, is 1 am = 931.494 MeV, which means:

$Q = 17.589$ MeV, an exothermic reaction

The total energy balance for this reaction is, by conservation of energy with the Tritium nucleus at rest, is, with kinetic energy leveraging classical physics:

$$Q + (T_d = \frac{1}{2} m_d v_d^2) + 0 \rightarrow (T_n = \frac{1}{2} m_n v_n^2) + (T_{He} = \frac{1}{2} m_{He} v_{He}^2)$$

In reality, for the reaction to occur, the deuterium (d) must overcome the Coulomb barrier of the Triton (T) in order to collide with a kinetic energy sufficient so that the nuclear strong force subsequently enables the compound nucleus formation of $(M_d + M_T)$; however, once within the range of the nuclear strong force, the reaction is exothermic, as denoted from the Q-value.

Therefore, if we consider the deuterium to be at rest, by conservation of linear momentum, assuming that $v_d \to 0$, we have:

$$\cancel{m_d v_d} 0 + 0 \to m_n v_n + m_{He} v_{He} \quad \text{which leads to} \quad \left(v_{He} = \frac{m_n v_n}{m_{He}} \right)$$

Which, applying this result to the above conservation of energy, leads to:

$$v_n^2 = \frac{2Q}{\left(m_n + \frac{m_n^2}{m_{He}} \right)} \frac{MeV}{amu}$$

Substituting values, this leads to $v_n^2 = 5.3 \, \text{MeV/amu}$, and since

$$T_n = \frac{1}{2} m_n v_n^2 = \frac{1}{2} (1.008665 \, amu) \left(5.3 \frac{MeV}{amu} \right) = 14.1 \, MeV$$

Which also means the helium recoil nucleus is $17.6 - 14.1 = \sim 3.5$ MeV.

Conservation principles with Q-values can be applied to any reaction, and are always applicable.

Text Organization

The remainder of the text contains a number of topics as follows:

Chapter 2 discusses probability and applicable laws of probability, focusing on those typically encountered in nuclear related subjects.

Chapter 3 introduces numerical computations; it is only an introduction, although the material covered serves as a lead in for an advanced course in computational numerical analysis.

Chapter 4 briefly covers complex numbers for completeness.

Chapter 5 reviews solution methods for ordinary differential equations, and emphasizes skills needed for flexibility in solving general problems.

Chapter 6 presents power series solution methods, and how these solutions are consistent with alternative methods.

Chapter 7 presents the solution methods needed for variable differential equations.

Chapter 8 covers vector, matrices, and linear systems solution methods with sufficient detail for accomplishing solution approaches.

Chapter 9 discusses Gram–Schmidt orthogonalization and Fourier Series, leading up to solutions of partial differential equations.

Chapter 10 presents "Applied Solution Methods" Part 1, and covers topics essential for preparation involving multiple dimensional applications.

Chapter 11 presents "Applied Solution Methods" Part 2, and explores applications in ordinary and partial differential equations, with a focus on the heat equation—the simplest PDE. Separation of Variables, superposition, and eigenfunction expansion are covered in the discussion.

Chapter 12 discusses introduces "neutronics"—briefly introducing neutron transport, moving into a discussion on neutron diffusion theory, with applications.

Chapter 13 introduces numerical solution methods of Partial Differential Equations (PDEs); this material also serves as a lead-in for an advanced course in computational numerical analysis.

A detailed set of application problem sets are included in the Appendix.

The depth and breadth of topics presented in each chapter varies; the goal is that, at the close of a course that covers the material presented in this text, a student in nuclear engineering should be ready to tackle the rigors of neutron transport theory, reactor physics, radiation interactions, radiation shielding, and other challenging topics in a nuclear engineering curriculum.

Chapter 2

Essentials of Probability

The study of probability began in the 17th Century as a result of a desire to determine the outcome of games of chance. By definition, a probability is a real number bounded between [0,1]. Often, probabilities are expressed as percentages, or positive outcomes in a finite number of trials, e.g. your odds of winning are "1 in 3." More formally:

Probability

Probability is defined as the mathematical modeling of the phenomenon of *chance* or *randomness*. A probability is always stated in a fraction: a real number from [0.0,1.0].

A complete study in probability begins with a discussion of set theory, elements, subsets, Venn Diagrams, set based proofs, association laws, etc. These are important topics, but are beyond the scope of our discussion here, where we provide a brief discussion of basic probability applications leading up to the likelihood of radioactive decay and probability distributions as they pertain to nuclear processes. It is from this perspective that we begin.

Counting Rules

There are two key rules of "event counting" that enable one to attribute specific events among a possible suite of events, and then determine a likelihood that the specific events occur. The two key rules are the "sum rule" and the "product rule."

Sum rule of counting

> Given: Event A occurs in m ways,
> Event B occurs in n ways....
>
> Then Event A ***or*** B can occur in $(m + n)$ ways.

Product rule of counting

> Given: Event A occurs in m ways, and independently,
> Event B occurs in n ways....

Then combinations of events A ***and*** B can occur in $(m \cdot n)$ ways.

Example

Suppose a student at the university has the following choices (assuming no prerequisites):

> 3 History courses, 4 English courses, 2 Science courses
>
> If a student must choose one of each, the number of ways in which he/she may do this is:
>
> $$(3)(4)(2) = 24$$
>
> If a student must choose only one course, the number of choices he/she has is:
>
> $$(3+4+2) = 9$$

The concept of an individual outcome for a number of trials is frequently encountered. In this context, consider the following:

Tossing a coin once.

Choosing a card from a deck.

Picking winning lotto numbers.

We note that while each individual event among those listed above has a specific outcome not explicitly known, an overall result will be known for "many trials…" Consider x and y as events.

Consider 2 events (or outcomes) x, y in N trials. After each trial 1, 2,…N, there are *four possibilities* for the outcome:

# of Times

x occurred but not y…	n_1
y occurred but not x…	n_2
x and y both occurred…	n_3
neither x nor y occurred…	n_4

The number of recorded events are: $(n_1 + n_2 + n_3 + n_4) = N$

Event Probability Metrics

We can define the probability for a number of events, noting that each is a probability that results in a number in the interval [0,1]:

Summary of Probability Rules: *Probability Set*

$$\frac{n_1 + n_3}{N} \leftrightarrow P(x) \equiv \text{probability } x \text{ occurred} \qquad A$$

$$\frac{n_2 + n_3}{N} \leftrightarrow P(y) \equiv \text{probability } y \text{ occurred} \qquad\qquad B$$

$$\frac{n_1 + n_2 + n_3}{N} \leftrightarrow P(x+y) \equiv \text{prob either } x \text{ or } y \text{ occurred} \qquad A \cup B$$

$$\frac{n_3}{N} \leftrightarrow P(xy) \equiv \text{probability both } x \text{ and } y \text{ occurred} \qquad A \cap B$$

Conditional Probability Rules   *Probability Set*

$$\frac{n_3}{n_2 + n_3} \leftrightarrow P(x \mid y) \equiv \text{probability of event } x, \text{ given} \qquad A \text{ given } B$$

$$\text{that } y \text{ has occurred}$$

$$\frac{n_3}{n_1 + n_3} \leftrightarrow P(y \mid x) \equiv \text{probability of event } y, \text{ given} \qquad B \text{ given } A$$

$$\text{that } x \text{ has occurred}$$

Mutual Exclusivity and "Complement"

"$A \cap B = \phi$" is known as "disjoint" or "mutually exclusive", where A and B cannot occur simultaneously.

The "Complement" involves the probability that the event does not occur, so that, for an event x with a probability $P(x)$, then:

$$P(x^c) = 1 - P(x)$$

Multiplication and Addition Laws

One can often use keywords "either" or "both/all" to distinguish what course of action to follow when determining probabilities. With the probability rules defined above, by inspection, we can write the multiplication and addition laws. Given two events x and y, the probability that *both events will occur* can be gleaned from the multiplication law, where probability sets intersect:

Multiplication Law of probability

$$P(xy) = P(x)P(y \mid x) = P(y)P(x \mid y)$$
$$A \cap B \text{ Implies "both" or "all"}$$

If we consider that *either of two events can occur*, the addition law applies, but requires that we account for the exclusion of "both" or "all" as determined from the multiplication law:

Addition Law of probability

$$P(x + y) = P(x) + P(y) - P(xy)$$
$$A \cup B \text{ Implies "either"}$$

A number of useful examples come from a deck of cards. In a standard deck of 52 cards, there are 12 face cards {Jacks - J, Queens - Q, Kings - K} spread among the 4 suits. Suits are {Diamonds - D, Hearts - H, Clubs - C, Spades - S}. There are then 13 cards in each suit, spanning "2" through "Ace."

Example

What is the probability of drawing one card that is the ace of spades? It is given there are 52 cards in the deck. Also x = drawing ace of spades = y, then each deck has only 1 ace of spades.

$$P(x) = P(y) = \frac{1}{52} \approx 0.019$$

Example

What is the probability of choosing the ace of spades twice in two independent draws?

Note that choosing the "ace of spades twice" in two independent draws is as follows: by multiplication law: $P(xy) = P(x)P(y \mid x)$ probability that the 1st draw is the ace of spades given that second draw is the ace of spades, and $P(x) = 1/52$ (where each draw is independent):

$$P(y \mid x) \rightarrow P(y) \quad \therefore P(xy) = \left(\frac{1}{52}\right)\left(\frac{1}{52}\right) \cong 0.00037$$

$\therefore$ it is unlikely that two consecutive random draws will yield the ace of spades twice in a row.

The "Complement" of drawing the Ace of Spades 2 times in a row is then

$$P((xy)^c) = 1 - 0.00037 = 0.99963$$

Example

For a standard deck, compute the probability that a

event A = a Heart   <u>and</u>
event B = a Face card   will be drawn.

Note: "and" implies the *Multiplication Rule*:

$$P(A) \equiv \text{probability we will draw a Heart} = \frac{13}{52} = \frac{1}{4}$$

$$P(B) \equiv \text{probability we will draw a face card} = \frac{12}{52} = \frac{3}{13}$$

$P(AB) = P(A) P(B)$ is also written

$$\therefore P(A \cap B) = P(A) P(B) = \frac{3}{52} = 0.0577$$

Example

Compute the probability that either a *heart* or a *face card* will be drawn from a standard deck.

Note: "Either" implies the *Addition Rule*

$$P(A) \equiv \text{probability we will draw a heart} = \frac{13}{52}$$

$$P(B) \equiv \text{probability we will draw a face card} = \frac{12}{52}$$

$P(A + B) = P(A) + P(B) - P(AB)$ is also written

$$P(A \cup B) = P(A) + P(B) - P(A \cap B)$$

$$\therefore P(A \cup B) = \frac{13}{52} + \frac{12}{52} - \left(\frac{13}{52}\right)\left(\frac{12}{52}\right)$$

where $P(A \cap B) \leftrightarrow P(AB) = P(A)P(B|A)$, each is independent

$$P(A \cap B) = 0.42307 \qquad \therefore P(AB) = P(A)P(B)$$

The Law of Large Numbers

Assume an experiment is repeated N times and "a" out of "n" times, the result was of type "x".

$$P(x) = \lim_{N \to \infty} \frac{a}{N} \Rightarrow \text{in practical limit, an infinite number of trials is not possible}$$

$$P(x) = \frac{a}{N} \quad \text{where } a/N \text{ is the frequency of occurrence of } x \text{ in the first } N \text{ trials}$$

Example

Consider a single coin toss

100 times $\rightarrow$ 47 heads

                        53 tails

Then $P_{Heads} = \dfrac{47}{100} = 0.47$

$P_{Tails} = \dfrac{53}{100} = 0.53$

As $N \rightarrow \infty$, the *law of large numbers* for N says that probabilities will yield their true theoretical outcomes... $P_{Heads} = P_{Tails} = 0.50$

Graphical Definition of Probability

Consider the following graphical definition of probability:

$$P(A) \equiv \frac{\# \ Pts \ in \ A}{\# \ Pts \ in \ S} \qquad\qquad P(S) = 1$$

$$\frac{f(A)}{n} \equiv \frac{\# \ times \ A \ occurs}{\# \ trials} \qquad where \ 0 \leq f(A) \leq 1$$

Note that here a "universe" is defined by the region "*S*," where the probability of a point being in "S" is 100% (with a $P((S)^c) = 0$). The region "*A*" is contained inside "S".

One can consider the number of points sampled randomly in "*S*," and with the Law of Large Numbers, ultimately $P(A)$ is the ratio of the area of "*A*" divided by the area of "*S*."

NUCLEAR APPLICATION: The Monte Carlo Method

The Monte Carlo method was named as such since it involves "playing the probability game"... sampling probabilities for a large number of trials (thus applying the Law of Large Numbers) to statistically arrive at a correct solution. The Monte Carlo method is simple in principal, but becomes fairly tedious without the use of a computer. A simple application Monte Carlo can be demonstrated by *Numerical Integration*, where a simple application of Monte Carlo is to "throw darts" in a closed area where a function is plotted, and determine the integral (area under the function) by using a simple ratio of the "darts" striking under the curve to the "darts" hitting the closed area.

Example

Consider estimating $\int_0^1 x^2 dx$ over a 1×1 square area. We "throw" random "darts" at the page within the sampling area over which the function is projected:

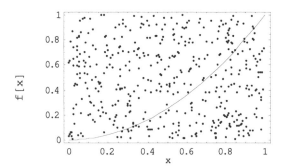

Representation of Monte Carlo Sampling of an Area

and 141/450 dots fell under the curve, estimating the integral to be 141/450*(Area) = 141/450 (1)(1) = 0.31333, where the actual answer is

$$\int_0^1 x^2 dx = 0.33333$$

...which can be better approximated by throwing more darts!

This points out that the *Monte Carlo method* is ALWAYS subject to *some statistical sampling error*, which in this case, is 6%.

The Monte Carlo method is used a great deal in nuclear engineering due to the complex issues involved in solving radiation transport problems, and it is indeed a valuable method. However, nuclear engineers should always question Monte Carlo results that do not include statistical sampling error, such as the 6% error noted in our simple numerical integration example. More theory is required in discussing the convergence and variance in Monte Carlo methods. In general applications, mean (μ) and variance (σ) are determined as follows:

$$\mu = E(x) = \frac{\int_{-\infty}^{\infty} x f(x)dx}{\int_{-\infty}^{\infty} f(x)dx} \quad E(x^2) = \frac{\int_{-\infty}^{\infty} x^2 f(x)dx}{\int_{-\infty}^{\infty} f(x)dx} \quad \sigma^2 = E(x^2) - \mu^2$$

Sampling With and Without Replacement

"Replacement," or a lack of, accounts for a shift in the relative probability as sampling occurs, and this is best discussed using an example.

Example

Consider that 3 screws are drawn from a lot of 100 that contain 10 defective ones…determine the probability that all 3 are non-defective (with replacement) if we can draw one that is not defective with a probability of 0.9.

With Replacement, 3 sequential draws become

$$(0.9)(0.9)(0.9) = 0.7290 \quad 72.9\%$$

Next, determine the probability that all 3 are non-defective *without replacement*

$$\left(\frac{90}{100}\right)\left(\frac{89}{99}\right)\left(\frac{88}{98}\right) = 0.7265 \quad 72.65\%$$

Permutations [n!]

A number of possible arrangements for a given set of *n* items is *n*!

Example

Consider a, b, c (a set of 3 letters of the alphabet).

The number of possible arrangements is:

abc, acb, bac, bca, cab, cba

the number of which can be determined from 3! = 3*2*1 = 6

Combinations

The combination formula can be used to determine the number of possible samples from a defined set. This then can be used to directly determine sampling probabilities.

The Combination formula is:

$$\binom{n}{k} \equiv \left[\frac{n!}{k!(n-k)!}\right] = \frac{n(n-1)...(n-k+1)}{1 \cdot 2 ... k}$$

and yields number of samples or combinations of "*n*" things "*k*" at a time.

Example

What is the probability of winning a "pick 6" lotto where 6 random numbers are chosen from $\{1...49\}$?

NOTE: let event $A \equiv WIN\ LOTTO$ from 1 ticket.

First, compute the number of combinations available from choosing 6 random numbers from the set of numbers contained in $\{1..49\}$:

$$\binom{49}{6} = \left[\frac{49!}{6!(49-6)!} \right] = 13.98 \times 10^6$$

Then, the odds of choosing one "winning combination" is

$$P(A) = \frac{1}{\binom{49}{6}} = \frac{1}{13.98 \times 10^6}$$

Example

What if you sold your car and bought $10k in unique lotto tickets? What is your new probability of choosing a winning ticket?

$$P(A \cup B) = P(x+y) = (10000)\frac{1}{13.98 \times 10^6} = 0.00072$$

(Note the Addition Rule for mutually exclusive events.)

Trying to win the lotto with realistic odds is not an easy (or likely a profitable) task!

Example

Two cards are drawn at random from a standard deck of 52 cards (4 suits of 2... 10, J, Q, K, A). Using the *Combination formula*, find the probability that both cards are Hearts.

Let event A ≡ Both Cards are Hearts.

First, we determine how many ways there are to choose *any* two cards using the combination formula:

$$\binom{52}{2} = 1326 \ \text{ways to choose 2 cards from the 52 card deck}$$

Then, we determine how many ways there are to choose two cards that are both Hearts:

$$\binom{13}{2} = 78 \ \text{ways to draw 2 Hearts from the 13 available}$$

The solution we seek comes from the **ratio** of these two combination formulas:

$$P(A) \approx \frac{\binom{13}{2}}{\binom{52}{2}} = \frac{78}{1326} \approx 0.058824$$

Example

Using the *Combination formula*, find the probability that one card is a

Heart and one Card is a Spade.

Let B ≡ one card is a Heart, and one card is a Spade.

As before:

$$\binom{52}{2} = 1326 \text{ ways to choose 2 cards from the 52 card deck}$$

Then, there are 13 hearts, and there are 13 spades; therefore, there are $(13)(13) = 169$ ways to draw a single heart and a single spade.

Note this is the same as computing

$$\binom{13}{1}\binom{13}{1} = (13)(13) = 169$$

And the solution is:

$$P(B) = \frac{(13)(13)}{\binom{52}{2}} = \frac{169}{1326} \approx 0.12745$$

Binomial Coefficients

The numbers from the combination formula

$\binom{n}{k}$ are called *Binomial Coefficients*, since they appear as the coefficients in the expansion of $(a+b)^n$.

Recall that $\binom{n}{k} \equiv \left[\dfrac{n!}{k!(n-k)!}\right]$

$$(a+b)^n = \sum_{k=0}^{n}\binom{n}{k}a^{n-k}b^{k}$$

$$(a+b)^1 = \sum_{k=0}^{1}\binom{1}{k}a^{1-k}b^k = \binom{1}{0}a^1b^0 + \binom{1}{1}a^0b^1 = (a+b)^1$$

$$\text{and } \binom{1}{0} = \frac{1!}{0!\,1!} \quad \binom{1}{1} = \frac{1!}{1!\,0!}$$

Pascal's Triangle

The *Binomial Coefficients* can be more easily derived using *Pascal's Triangle*, shown below. Note the first and last number of each row is 1, and elements are derived from the sum of adjacent elements above.

$$
\begin{array}{ccccccccccc}
 & & & & & 1 & & & & & \\
 & & & & 1 & & 1 & & & & \\
 & & & 1 & & 2 & & 1 & & & \\
 & & 1 & & 3 & & 3 & & 1 & & \\
 & 1 & & 4 & & 6 & & 4 & & 1 & \\
1 & & 5 & & 10 & & 10 & & 5 & & 1 \\
 & \cdots & & & & & & & & &
\end{array}
$$

$$\textit{Pascal's Triangle}$$

e.g. consider the following:

$$(a+b)^3 = a^3 + 3a^2b + 3ab^2 + b^3$$

NUCLEAR APPLICATION: Nuclear System Reliability

Example

Consider two redundant, hypothetical nuclear rated valves at a nuclear power plant, where we have valve #1 on cooling circuit #1, and valve #2 on cooling circuit #2. Then consider the following:

event A = valve #1 will <u>operate</u> for 20 years without failure *event B* = valve #2 will <u>operate</u> for 20 years without failure

We will assume A and B are independent events, and that

$$P(A) = \frac{1}{4} \qquad P(B) = \frac{1}{3}$$

(note these values are orders of magnitude higher than one would encounter for illustrative purposes only!)

The probability that <u>both</u> will operate after 20 years is given by the multiplication rule (note this is indicated by the "both" statement), where:

$P(AB) = P(A)\,P(B)$ is also written

$$P(A \cap B) = P(A)P(B) = \frac{1}{12} = 0.0833$$

The probability that either *A* or *B* will operate after 20 years, as suggested by the wording, is governed by the addition rule, where

$P(A + B) = P(A) + P(B) - P(AB)$ is also written

$$P(A \cup B) = P(A) + P(B) - P(A \cap B)$$

$$P(A \cup B) = \frac{1}{4} + \frac{1}{3} - \frac{1}{12} = \frac{1}{2}$$

The probability that neither *A* or *B* will operate after 20 years is therefore the complement of $P(A \cup B)$:

$$P((A \cup B)^c) = 1 - P(A \cup B) = \frac{1}{2}$$

Since these events are independent, this could also be determined from a basis using complements, with the probabilities the valves will *fail*:

$$P((A)^c) = \frac{3}{4} \qquad\qquad P((B)^c) = \frac{2}{3}$$

Then

$$P((A)^c \cap (B)^c) = \frac{3}{4} \cdot \frac{2}{3} = \frac{1}{2}$$

The probability that only valve B will be operating after 20 years is therefore:

$$P((A)^c \cap B) = \frac{3}{4} \cdot \frac{1}{3} = \frac{1}{4}$$

The probability of survival or failure pertaining to equipment in a nuclear power plant is of course a major issue in licensing the facility for safe operation, and numerous redundancies must be analyzed to assess the risks to the plant.

Note here we have treated the two events considered to be independent. While this is normally the case, it is not always true, and interdependent probabilities can become quite complex to evaluate. Often, it is simpler to have independent redundant systems, since this tends to reduce failure and simplifies the analysis.

Finally, large diagrams that resemble "trees" map out specific chains of events, and this procedure is called "fault tree analysis", or FTA. FTA is a "top-down" approach for analyzing potential system failures before they occur for systems under development, beginning with the top event (the initiating potential failure). Analysts then determine all pathways the failure leads to that affect the system, which ultimately determines risk.

This is a routine practice in engineering design, particularly important in nuclear engineering for obvious reasons.

NUCLEAR APPLICATION: Radioactive Decay

An atom with an unstable nucleus will, at some point in time, rearrange itself to a more stable configuration, although the precise instance of when this will occur is, for an individual atom, *a low probability event.*

For example, consider the decay probability of a Pu-239 nucleus per second, 9.114E-13 1/s.

If a significant number of like unstable isotopes can be gathered together, then the rearrangements, or "decays" can be made at a predictable rate that is individually characteristic of the isotope.

The "decay" process is traditionally measured in terms of a "half life" where ½ of the nuclides decay after a measure of time:

$$\frac{dN}{dt} = -\lambda N$$

Recognizing this is a separable differential equation describing $N(t)$, then integrating yields $\ln(N) = -\lambda t + c_1$

Taking an exponential of both sides...

$$N(t) = \exp(c_1)\exp(-\lambda t)$$

and assuming we start from an initial number N_o, then

$$N(t) = N_o \exp(-\lambda t)$$

The probability of decay of a nuclide λ, in the interval between t and $t + dt$, even if small, can be determined from the half life measured from a very large number of nuclei decaying (applying the Law of Large Numbers):

$$\frac{N(T_{half})}{N_o} = \frac{1}{2} = \exp(-\lambda T_{half})$$

Then, taking the log of both sides, canceling signs, and solving for the probability: $\lambda = \ln(2)/T^{1/2}$. Often one must consider a practical limit for when an isotope has "disappeared," even though it is never fully "gone." Usually, this is after 5 to 10 half lives, depending on limits, where one part in 32 (~3%) or one part in 1024 (~0.1%) remain, respectively.

In the early years of nuclear physics and engineering, radiochemists determined half lives of isotopes graphically by using the fact that two nuclides with different half lives will decay away at different rates. By plotting the relative activities of the combined "cocktail" of nuclides and iterating on initial guesses for half lives based on the graph, the half lives of the two decaying isotopes can be isolated, as shown below.

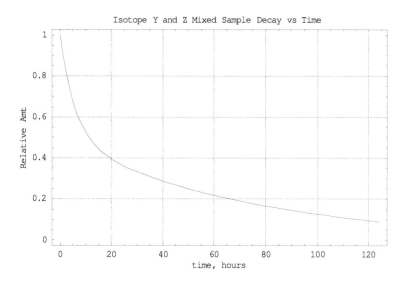

Two isotopes with half lives of 4 and 50 hours, respectively

Additional Decay Exercises

Many nuclides form and decay in a time too short to measure, such as:
U-238 + n → U239* → U-239 + gamma in ~1.E-14 s,
where others take longer, as in free neutron decay:

$\quad$ n → p + β after ~12 min

Each of these events has a probability.

(i) Show that the probability a nuclide will survive up to a time t, then decay in a time $t + dt$ is: $p(t)dt = \text{Exp}[-\lambda\, t]\lambda dt$

(ii) Integrate this expression from $[0, \infty]$ (over all time from this point forward) to show that the probability is 1.0.

(iii) Demonstrate that the mean life of the nuclide is $1/\lambda$

NUCLEAR APPLICATION: *Probability Distributions and the Fission Spectrum*

Fission neutrons are emitted over a continuous probability distribution of energy, known in nuclear engineering as the *Prompt Fission Neutron Spectrum*.... Note this implies the existence of a *Delayed Neutron Spectrum.* A brief word on the subject of *prompt* versus *delayed* neutrons is worth noting. Most neutrons emitted in fission are in fact prompt—they are emitted within 1E-13 seconds of a fission event. Indeed, a small fraction of the neutrons emitted in fission are *delayed* in their emission following a fission event, with half-lives measured as long as *several minutes.* This small fraction of neutrons, emitted at some protracted time period(s) after a fission event, is known as β—the delayed neutron fraction, and it is 0.65% for U-235, 0.21% for Pu-239. *Were it not for the delayed neutron fraction*, the only nuclear power available to us would be the *uncontrolled chain reaction of nuclear weapons*, since the delayed neutrons are emitted in times reasonable for mechanical control of the fission chain reaction in a reactor. Needless to say, when a reactor does not rely on the delayed neutrons for a chain reaction, this is described as being in a "*prompt critical*" condition. Unless the reactor is designed for this in a "*pulse*" condition, such a condition would normally result in a reactor accident; many safeguards are in place to prevent "prompt" criticality. We will return to this subject a bit later.

Now, consider that $\chi(E)$ is called the *Prompt Fission Neutron Spectrum.* This spectrum for U-235 is given below:

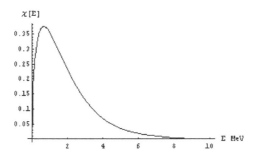

Prompt fission spectrum for U-235

For a U^{235} nucleus, the Maxwellian Prompt Fission Neutron Spectrum is:

$$\left[\chi(E) = \frac{77}{100}\sqrt{E}e^{-\left(\frac{0.776}{MeV}E\right)}\right] \text{ with } E \text{ in } MeV$$

$\chi(E)\,dE \Rightarrow$ *is the average number of neutrons emitted with an energy E between E and E + dE per fission neutron,*

where $dE = \lim_{\Delta E \to 0} \Delta E$

Another way to think of this is:

$\chi(E)\,dE \Rightarrow$ *is the net fraction of fission neutrons emitted with an energy E between E and E + dE.*

Note that Energy in Nuclear Engineering is given in units of (eV, KeV, or MeV), 1 electron volt (eV) is the energy imparted to a free electron when accelerated by an electric field potential of 1 Volt. Specifically,

$$1\ eV = 1.602 \times 10^{-19}\ J \text{ and } 1\ MeV = 1.602 \times 10^{-13}\ J$$

Noting that by definition, $\displaystyle\int_0^\infty \chi(E)dE = 1$

$$\Rightarrow \chi(E) = C_m \sqrt{E}e^{-\alpha E} \text{ where } C_m \equiv 0.77, \quad \alpha \equiv 0.776$$

Now solve for the most probable fission neutron energy:

$$\frac{d\chi}{dE} = \left[\frac{77}{100}\frac{1}{2}E^{-\frac{1}{2}}e^{-0.776E} + \frac{77}{100}E^{\frac{1}{2}}e^{(-0.776E)}(-0.776)\right]$$

$$\text{Set } \frac{d\chi}{dE} = \frac{77}{100} e^{-0.776E} \left(\frac{1}{2\sqrt{E}} + E^{1/2}(-0.776) \right) = 0$$

$$\frac{1}{2\sqrt{E}} + \sqrt{E}(-0.776) = 0 \text{ or } \frac{1}{2\sqrt{E}} = \sqrt{E}(0.776)$$

$$\text{Then } E = \frac{1}{2(0.776)} = 0.6443$$

Typically, the fission spectrum is given as either a *Maxwellian Fission Spectrum* or a *Watt Fission Spectrum*. The forms are:

Maxwellian Spectrum: $\chi(E) = C_m E^{\frac{1}{2}} e^{-\alpha E}$

Watt Spectrum: $\chi(E) = C_w \exp(-\alpha E) \sinh \sqrt{(\beta E)}$

Note that a unique $\chi(E)$ exists for every fissile nuclide. It is therefore always important to verify that

$$\int_0^\infty \chi(E) dE = 1$$

for the spectrum given. The *average fission neutron energy* is computed from

$$\overline{E} = \frac{\int_0^\infty E\chi(E) dE}{\int_0^\infty \chi(E) dE} = \frac{1}{1} \int_0^\infty E\chi(E) dE$$

Really, this is an "improper integral"

$$\lim_{b \to \infty} \int_0^b E(C_m E^{1/2} e^{-\alpha E}) dE = \lim_{b \to \infty} \int_0^b C_m E^{3/2} e^{-\alpha E} dE$$

This is a special form of the exponential integral

$$\lim_{b \to \infty} \int_0^b C_m x^{3/2} e^{-\alpha x} dx$$

$$\lim_{b \to \infty} \left[e^{-\alpha x} C_m \left(-\frac{3\sqrt{x}}{2\alpha^2} - \frac{x^{3/2}}{\alpha} \right) + \frac{3\sqrt{\pi} C_m \, Erf\left[\sqrt{\alpha \, x}\right]}{4(\alpha)^{5/2}} \right]_0^b$$

Where *Erf* [*x*] is the Gaussian Error Function, given below.

Gaussian Error Function

$$Erf[x] = \frac{2}{\sqrt{\pi}} \int_0^x \exp[-u^2] du \quad \text{with}$$

$$\lim_{b \to 0} Erf[b] = 0 \quad \text{and} \quad \lim_{b \to \infty} Erf[b] = 1$$

Standard Probability Distributions

Binomial Distribution

The *Binomial Distribution* is a probability distribution function (pdf) that can be applied in the following cases:

There are two possible outcomes: {A, B}

P(A) is constant, independent of no. of observations and P(B) = 1 − P(A); Occurrences of A observed (or B observed) do not affect P(A) or P(B).

$$f_B(x) = \binom{n}{x} p^x q^{n-x} = \binom{n}{x} p^x (1-p)^{n-x}$$

So that the Binomial Distribution Formula is

$$f_B(x) = \left[\frac{n!}{x!(n-x)!} p^x q^{n-x} \right]$$

The Binomial Distribution Formula applies in cases like a *coin toss* and in the *decay of a few atoms*.

Example

What is the probability of getting heads in a coin toss three times in a row?

$A \equiv$ get a heads up coin toss
$B \equiv$ don't get a heads up coin toss (tails)
$n = 3$ total trials

$x =$ getting heads 3 times in a row... $p = 0.5$ $q = (1 - 0.5)$

$$f_B(3) = \left[\frac{3!}{3!(3-3)!} 0.5^3 (1-0.5)^{3-3} \right] = 0.125$$

Two other important Probability Distribution Functions (PDFs) include the *Poisson Distribution* and the *Normal Distribution*.

Poisson Distribution

This applies to events with a *small* but *constant* probability of occurrence, and is derived from Binomial Probability Density Function, we apply the limits as

$$\begin{Bmatrix} n \to \infty \\ p \to 0 \end{Bmatrix} \text{ so that } f_p(x) = \frac{\mu^x}{x!} e^{-\mu}$$

where $\mu \equiv$ average ("mean") for a <u>large number</u> of trials, and $x \equiv$ the result of the very next trial

$$\text{and } \sigma_p = \sqrt{\mu}$$

NUCLEAR APPLICATION: Radiation Detection Statistics

A radiation detector is used to count particles from a radioactive isotope. If the average count rate is 20 counts/min (cpm), what is the probability the next trial will yield 18 cpm?

Then: $\mu = 20$, $x = 18$ $f_p(x) = \dfrac{20^{18}}{18!} e^{-20} = 0.0844$ or $\sim 8\%$

NOTE: As the mean *increases*, the Poisson distribution becomes *symmetric about the mean.*

Other important things to remember:

$\to$ Both the *Binomial* and *Poisson* distributions apply to <u>discrete variables or events</u>.

$\to$ Most variables in experiments are continuous...

Gaussian Distribution

$$f_g(x) = \frac{1}{\sqrt{2\pi}\sigma} \exp\left[-\frac{1}{2}\left(\frac{x-\mu}{\sigma}\right)^2\right]$$

about "μ" (the mean), and σ is the standard deviation,

where x extends to $-\infty < x < \infty$

The *Full Width Half-Maximum* (FWHM) of the Gaussian function is defined as the peak width at precisely one half of the peak maximum, and is determined mathematically by:

$$f_g\left(\mu - \frac{\Gamma}{2}\right) = f_g\left(\mu + \frac{\Gamma}{2}\right) = \frac{1}{2}f_g(\mu)$$

Solving this yields the relationship between the FWHM and the uncertainty: $\Gamma = 2.35\sigma$.

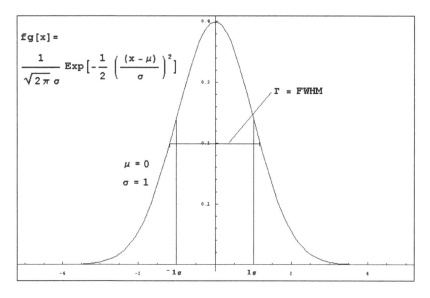

A normalized Gaussian distribution function with a zero mean, unit standard deviation.

Central Limit Theorem

The *Central Limit Theorem* states that sampled mean values, e.g. from a collection of Monte Carlo sampling histories, follow a Gaussian distribution for n stochastic samples of a probability density function, and that the *uncertainty is proportional to the inverse of the square root of the number of particle histories.*

This is a foundational principle in applying Monte Carlo algorithms to nuclear problems.

NUCLEAR APPLICATION: Probability in Medical Physics

Example

You are asked to run a Monte Carlo simulation to estimate a dose that will be delivered to a patient, and your supervisor has asked for a solution with 1.5% or less uncertainty.

Assume the *Central Limit Theorem* can be applied. The Chief of Physics calls to tell you the solution is needed as quickly as possible for a high risk treatment plan. He tells you he is waiting by the phone for your answer.

You decide to set up an initial one hour run of the Monte Carlo simulation, and after 60 minutes of computing time, the run completed $n = 1,000,000$ particle histories, and the dose tally of interest reported by the Monte Carlo code is 170 Gy/hr $+/-$ 0.4%.

Since you only needed 1.5% accuracy and the Chief of Physics was waiting, estimate how many fewer histories could have been run, and in how much time a shorter run would have required to achieve tally results within $+/-$ 1.5%.

To solve this, we assume that the central limit theorem applies, so that we can assume that the standard deviation is proportional to the inverse of the square root of the number of histories sampled:

$$\sigma = k / \sqrt{N}$$

From this, for the example given, we determine that $k = 400$. With this knowledge, we can then apply the same principle, assigning $\sigma = 1.5$ 1 and solving for a new N, which at this point yields 71,111 histories required for the standard deviation given.

Then, assuming a linear extrapolation of histories sampled correctly accounts for computation time (a reasonable assumption), we obtain that the actual model should have only required less than 4.3 minutes to yield a 1.5% accuracy goal, and that 60 minutes was quite an overestimate of the time.

NUCLEAR APPLICATION: Detector Peak Resolution

Detector Peak *Resolution* (R) and the FWHM is an often discussed issue when dealing with radiation detectors. Consider scintillator detectors attached to multi-channel analyzers, which "bin" the energy values initially recorded as light pulses as a result of interactions of the radiation in a gamma radiation scintillator detector. Due to the variation in the signals, these events yield Gaussian peaks that have a mean based on the energy of the incident gamma radiation.

Resolution of the detector is defined as a percentage derived from

$$R(\mu) = \frac{\Gamma}{\mu}$$

so that detectors with better resolution have narrow FWHMs at a given energy. For example, if the normalized Gaussian in the last figure depicted a counts vs energy over a reasonable number of channels for a detector, and if $\mu = 662$ keV for the gamma ray energy emitted from a Cs-137 gamma ray emitting isotope (Cesium-137 is a fission product), if the resolution of the detector at 662 keV was $R = 3\%$, then you would know that the FWHM at 662 keV was 19.9 keV.

Resolution is an important fact in being able to determine if one can clearly distinguish between two gamma lines in a gamma radiation detector, and it is desirable to have the FWHM as low as possible.

Integration of the Gaussian Function

Integration of the Gaussian yields a special function denoted as $Erf(x)$, the *Gaussian Error Function, or ERF:*

$$Erf(x) = \frac{2}{\sqrt{\pi}} \int_0^x \exp(-t^2)dt$$

$$A_\sigma = \int_{\mu-\sigma}^{\mu+\sigma} f_g(x)dx = 2\,Erf(\mu+\sigma) = 0.683$$

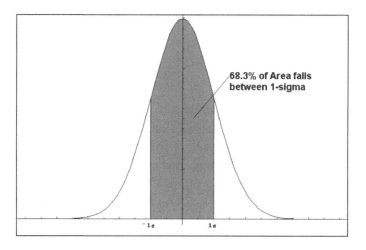

A graphical depiction of 68% of the area under the Gaussian function.

For a normalized Gaussian, 68.3% of the curve falls under the interval characterized by "1-sigma" spanning the mean. Similar values for "2-sigma" and "3-sigma" are 95.4% and 99.7% of the area, respectively. This is often coined as the "68-95-99+" rule. To yield precisely 95%, the integration limits are $\mp 1.96\sigma$. This will be discussed further in the section discussing the Currie Limit for minimum detection.

Gaussian Representation of Poisson

Consider again the two distributions, the Poisson and the Gaussian, where for purposes of discussion, we will treat the Poisson distribution as a continuous one, even though this is not a standard practice; therefore, the Poisson (fp) and Gaussian (fg) distributions, respectively, are written according to *Mathematica* notation:

$$\text{fp}[x\_,\mu\_] := \frac{\mu^x}{x!}\text{Exp}[-\mu]$$

$$\text{fg}[x\_,\mu\_,\sigma\_] := \frac{1}{\sqrt{2\pi}\sigma}\text{Exp}[-\frac{1}{2}(\frac{(x-\mu)}{\sigma})^2]$$

Then, using *Mathematica*, we plot the two distributions (with Gaussian in Grey, and $\sigma = \sqrt{\mu}$) for means that vary from 3 to 35 in steps of 12 generates the following plots as continuous distributions (using the "GraphicsGrid" command in *Mathematica*):

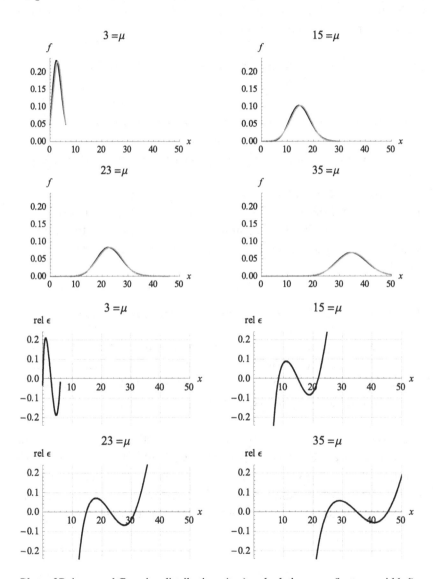

Plots of Poisson and Gaussian distributions (top) and relative error (bottom, gridded)

The plots of Poisson and Gaussian distributions in the figure were plotted for different mean values; it can be readily observed that as the mean (μ) value increases, the Poisson distributions become more symmetric about the mean, and the relative error between the two distributions becomes smaller relative to the mean center. At $\mu = 3$, there is $\sim 22\%$ maximum relative error computed between the two distributions.

The *Mathematica* command to generate these plots is as follows:

$$\text{Plot}[\{\text{fp}[x, \mu], \text{fg}[x, \mu, \sqrt{\mu}]\}, \{x, 0, 2\mu\}, \text{PlotRange} \rightarrow$$
$$\{\{0, 50\}, \{0, 0.24\}\}, \text{Frame} \rightarrow \text{False}, \text{AxesLabel} \rightarrow \{x, f\}, \text{PlotLabel} \rightarrow$$
$$" = \mu"\,\mu, \text{PlotStyle} \rightarrow$$
$$\{\{\text{Thickness}[0.009], \text{GrayLevel}[0]\}, \{\text{Thickness}[0.01]\text{GrayLevel}[0.55]\}\}], \{\mu, 3, 15, 12\}]$$

The convention is often taken that the smallest mean value of ~25 is the minimum value to be considered in any case where a Gaussian distribution (with $\sigma = \sqrt{\mu}$) can be used to approximate a Poisson distribution. At levels below this, Bayesian statistics (see a reference on statistics) must be applied.

NUCLEAR APPLICATION: *Currie Limit Detector Protocol*

To outline an approach to determine a minimum number of radiation detector counts, one can employ the traditional "Currie Detection Limit Protocol" formulation, assuming the number of radiation events in a detector follows a Gaussian distribution.

Generally, a Poisson driven process like radioactive decay can be well approximated by a Gaussian distribution when the mean is at least 25. Therefore, this is often set as the minimum number of total detector counts. However, there will often be enough of a variation in counts detected due to the statistical nature of the process, as mentioned previously.

To integrate a normalized Gaussian to yield 95% of the area, the limits needed are $\mp 1.96\sigma$ from the central mean (maximum). Based on the procedure outlined by Currie, the decision threshold regarding "did we or did we not detect something" will be established by, and is significantly dependent upon, the local detector background radiation signal.

Principally, we can assume that the total integrated counts come from a signal (that we wish to detect, such as radiation from Special Nuclear Material radiation sources (SNM), such as Pu) plus background.

Assuming the radiation signal is quite weak and very difficult to distinguish from background, the number of *detected counts* (D) is determined by subtracting the background (B) counts from the total counts collected (T); the variance associated with this scenario for the detected counts, by propagation of error is:

$$\sigma_D^{\ 2} = (\sigma_B^{\ 2} + \sigma_T^{\ 2}) \approx 2\sigma_B^{\ 2}$$

and with Poisson statistics, if operating for a detector Count time T_C in a background radiation count *rate* $\dot{B}$ we have $\sigma_B^{\ 2} = \dot{B}T_C$ and therefore the uncertainty in the detected counts can be expressed as:

$$\sigma_D \approx \sqrt{2\dot{B}T_C}$$

Where $\dot{B}$ = Background count rate, $\dot{D}$ = Detector net count rate, and T_C = Counting time.

The Currie Limit (L_C) is therefore established so that the minimum detectable activity for 5% probability of false alarm (PFA) (1.96σ above the mean) is:

$$L_C = 1.96\sqrt{2\dot{B}T_C}$$

where only up to 95% of the background area is accounted for. The PFA is the tare fraction of the background variance not accounted for that could be (in error) and associated with a positive detection.

Therefore, the Currie Limit (L_C) is depicted as the "Signal Decision Threshold Value" rate line in the following figure:

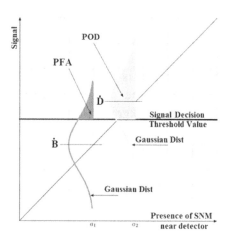

Illustration of Detector and Background count distributions; PFA is Probability of False Alarm, and POD is Probability of Detection

Then, assuming real activity that is truly present is indeed greater than this Currie Limit (L_C, the "Signal Decision Threshold"), the minimum detectable activity (N_D) for 95% POD/5% PFA, can then be computed as:

$$N_D = L_C + 1.96\sqrt{2\dot{B}T_C + \dot{D}T_C}$$

where L_C = Currie Limit count threshold value attributed to background variance

$\dot{B}$ = Background count rate

T_C = Counting time

$\dot{D}$ = Detector net count rate

N_D = Real Activity Present Threshold Counts

Example

Your criterion is 95%POD/5%PFA for a scenario where someone has stolen a radiation source and is driving it off site in a vehicle, and you are to use a roadside detector to try to find this source as vehicles drive by. In the geometry you will use at a toll booth checkpoint, you calculate that the source with a zero background will yield and average of $\dot{D} = 1$ count per second (cps). You will have a $T_C = 1$ s detector count time to work with since vehicles quickly pass by the detector. If the background rate is $\dot{B} = 1.72$ cps, determine the Currie limit and number of counts that will be a true positive during the counting time.

The Currie limit for 5% PFA is $1.96\sqrt{2\dot{B}T_C} \approx 3.64$ counts.

With 95% confidence in detection, a positive detection will yield

$$N_D = L_C + 1.96\sqrt{2\dot{B}T_C + \dot{D}T_C} \approx 7.8 \text{ counts.}$$

If the POD were lowered to 68% (1-sigma), then:

$$N_D = L_C + 1.00\sqrt{2\dot{B}T_C + \dot{D}T_C} \approx 5.75 \text{ counts}$$

Clearly, this example pushes the limits of "Poisson statistics" approximated using a Gaussian, and the lower the mean values measured, the more potential variation from predictable amounts will result. It is up to the user to apply limits to the statistical results to deem what is acceptable in each application. More advanced methods are available; the reader is encouraged to consult an advanced statistics reference.

This chapter provided a *very* brief look at probability and some issues related to it applied to problems of interest to nuclear engineers and medical physicists. It is recommended that the student further study this interesting and at times complex subject with a dedicated text on the subject.

Chapter 3

Introduction to Numerical Concepts and Applications

Most problems of practical interest in nuclear engineering and medical physics require the use of numerical algorithms to yield a solution. In fact, many of the quantities of interest can only be solved numerically. No matter what is being computed numerically, it is a fundamental truth that there are inherent assumptions, approximations, and limitations, and also *errors* associated with numerical computations. Therefore, it is important for one to have an appreciation for the issues involved in accomplishing numerical calculations.

The complexity of nuclear systems and applications is such that any approximations, issues, and potential numerical errors must be well understood by the engineer applying those methods. All numerical approximations involve incorporating a mathematical model. The following are steps to follow for *Numerical Problem Solving* that should be used to afford a numerical solution:

1. *Select* a mathematical *model*

2. *Evaluate* the model and associated limitations

3. *Implement* a programming solution

4. *Execute* computation on a computing system

5. *Evaluate* the solution and results

Models are just that—they are selected to represent the quantity or system under investigation, but they are limited by the detail included by the model developer to represent the true systems being used.

A programming solution is an algorithm, based on a model, that incorporates a series of logical solution steps, numerical assumptions, and implementations of some form of computer language to afford a solution. This could range from steps completed by hand, punched on a calculator, or as executed code on a multiprocessor computer. Either way, the programming solution is executed.

Precision and Accuracy

Results must be evaluated for both precision and accuracy. These were dealt with already in Chapter 2 in varying degrees. Again, precision can be associated with the amount of spread in data derived by repetitions of an experiment; the degree of precision is usually evaluated based on the variance in the answer. As discussed, precision can be improved by increasing the number of experiments.

Accuracy can be associated with how well a computed quantity relates to real world results; a model can be very precise but have poor accuracy. These concepts will be developed further as we present relevant examples.

In the next sections, we present the essentials of coordinate systems and other topics essential for numerical solutions typically encountered in solving nuclear engineering and medical physics problems.

Coordinate Systems and Differential Volume Elements

The three standard coordinate systems often applied in nuclear engineering problems include Cartesian, Cylindrical, and Spherical coordinate systems. More on how coordinate systems are derived will be discussed later. For now, the variables of each coordinate system and their respective differential volume elements are described below:

Cartesian Coordinate location:

x, y, z

Differential Volume:

$dV = dxdydz$

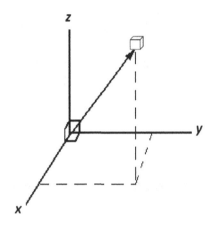

Figure Depicting Cartesian Coordinates with a differential volume element

Cylindrical Coordinate location:

r, θ, z

Differential Volume:

$dV = rd\theta\, drdz$

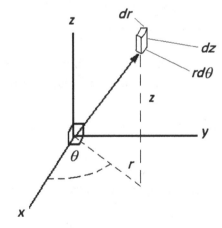

Figure Depicting Cylindrical Coordinates with a differential volume element

Spherical Coordinate location:
r, θ, φ

Differential Volume:
$dV = r \sin \theta d\varphi \ r \ d\theta \ dr$

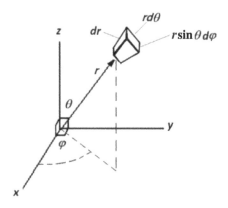

Figure Depicting Spherical Coordinates with a differential volume element

Applications in Spherical Coordinates: Solid Angle

Solid angle Ω is a measure of the angular region taken up by an object in space, equal to the area it subtends on a sphere divided by the radius of the sphere squared (r^2).

The solid angle is measured in *steradians* (sr), and falls between [0 and 4π]. The mathematical expression for a differential solid angle is

$$d\Omega = \frac{dA}{r^2}$$

Hence, in a spherical coordinate system, with a polar angle θ and an azimuthal angle ϕ, the differential solid angle is then

$$d\Omega = \frac{r \sin(\theta) d\varphi \ r d\theta}{r^2} = \sin(\theta) d\theta \ d\varphi$$

The solid angle is used often in applied radiation detection, where one must account for the detector efficiency, in part, as a result of the particles that may occupy the solid angle to impinge across the surface of a radiation detector.

Example

Consider a 10 cm diameter cylindrical detector placed 30 cm away from a point source of radiation, where the source emits gamma radiation isotropically (equiprobable in all directions). Determine the solid angle Ω_o of the detector for this geometry.

The solid angle subtended by the detector is computed using the following steps:

The detector radius is 5 cm, and the polar angle is computed from

$$\theta_o = \arctan(\frac{5cm}{30cm}) \approx 0.16515 \text{ radians}$$

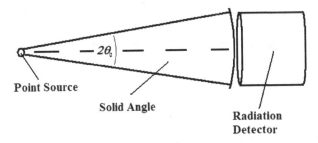

Figure: Geometry of Radiation detector and point source in determining solid angle

$$\Omega_o = \int\limits_{0}^{2\pi} \int\limits_{0}^{\theta_0} \sin(\theta)d\theta\, d\varphi \approx 0.0855 \text{ sr}$$

This means since the largest solid angle is 4π steradians, only about 0.7% of the radiation emanating from the point source impinges on the detector. An approximate method of computing the solid angle can be used if the area presented by the detector face is divided by the area of a sphere having a radius equal to the distance separating the source and the detector. The relative accuracy of this approximation for computing a solid angle improves as the source to detector range increases and detector area decreases, but can be in significant error as the source range to the detector decreases, or as the detector surface area presented to the source increases. Determining solid angle for more complex source and detector geometries can be quite challenging.

Composite Functions

Composite functions are functions that have variables, where these variables are also functions of *other variables*... Consider the following examples:

Example

$$z = f(x, y) \text{ with } x = g(t) \text{ and } y = h(t)$$

If f, g, and h are differentiable, from the chain rule, $\dfrac{dz}{dt}$ is:

$$\frac{dz}{dt} = \frac{\partial z}{\partial x}\frac{dx}{dt} + \frac{\partial z}{\partial y}\frac{dy}{dt}$$

Note that $\dfrac{dz}{dt}$ is a *total* derivative.

Example

$$z = f(x, y) \text{ with } x = g(t, u) \text{ and } y = h(t, u)$$

If f, g, and h are differentiable, then from the chain rule, $\dfrac{\partial z}{\partial t}$ is:

$$\frac{\partial z}{\partial t} = \frac{\partial z}{\partial x}\frac{\partial x}{\partial t} + \frac{\partial z}{\partial y}\frac{\partial y}{\partial t}$$

Note that $\frac{\partial z}{\partial t}$ is a *partial* derivative, and that since x and y are *both* *multivariate* functions, we must use *partial_derivatives* rather than *total* *derivatives* in the above expression.

Example

Given:

$$z = f(x, y) = x^2 + y^2 \quad with \quad x = 2t \quad and \quad y = \sin(t), \; find \; \frac{dz}{dt}$$

Again, from the chain rule:

$$\frac{dz}{dt} = \frac{\partial z}{\partial x}\frac{dx}{dt} + \frac{\partial z}{\partial y}\frac{dy}{dt}$$

Therefore, we compute each derivative:

$$\frac{\partial z}{\partial x} = 2x \qquad \frac{\partial z}{\partial y} = 2y$$

$$\frac{dx}{dt} = 2 \qquad \frac{dy}{dt} = \cos(t)$$

Putting it all together…

$$\frac{dz}{dt} = 2x \cdot 2 + 2y \cdot \cos(t)$$

Which then simplifies to:

$$\frac{dz}{dt} = 8t + 2\sin(t)\cos(t)$$

Note this gives the overall rate of change of "z" with respect to "t"!

Multidimensional Optimization

In order to optimize any computation for local extrema, particularly when subject to a constraint, can directly be solved using a gradient search using LaGrange multipliers.

Given an objective function

$$f(x, y, z)$$

and a constraint condition given by the function

$$g(x, y, z)$$

then this can be solved by applying the gradient and including the constraint equation, resulting in 4 equations and four unknowns:

$$\nabla f(x, y, z) - \lambda \nabla g(x, y, z) = \vec{0}$$

$$g(x, y, z) = 0$$

The parameter λ serves to "scale" the constraint equation to balance the objective equation, so that the critical values of x, y, and z can be determined directly.

Fundamental Concepts

There are also a number of fundamental concepts to consider, given below.

Significant Digits

The amount significant digits for a number α is defined as any digit to the right of and including the first non-zero digit...

Example

<u>1.6670</u>E-31   <u>16670.</u>   <u>1.6670</u>   0.000<u>16670</u>

All of the above numbers have five (5) significant digits.

Floating Point System

A floating point system us where the number of significant digits is fixed and the decimal "floats" to describe the value, such as presented below:

6.023E+23   5.67E-08

Sources of Error--

Underflow and overflow

Underflow or overflow are inherently violations of single or double precision for real numbers, as defined for computation by IEEE

IEEE single precision, REAL*4 FORTRAN,
designated as ($-38 <$ exponent < 38)

IEEE double precision, REAL*8 FORTRAN,
designated as ($-308 <$ exponent < 308)

Round off or "chopping"

Round off – rounding last digit
Chopping – dropping/truncating last digits

Example: "1.2535"

Rounds to…

Decimal 3 digit	Decimal 2 digit	Decimal 1 digit
1.254	1.25	1.3

Also, "1.25" rounds to 1.2 with no other specification.

Algorithm Stability

Small changes in initial data should yield "linear" or "bounded" changes in outcomes or results.

Algorithms that are extremely "stiff" (where large changes result from small changes in independent variables) are often "unstable" or "non-convergent".

There is significant theory dedicated to methodologies to employ in identifying stable, convergent algorithms versus unstable, non-convergent algorithms. Entire courses can be devoted to this subject in significant detail. We will touch on a few concepts.

Root Solving Methods

$f(x) = 0$ is denoted as a *root solving problem*

A *root solving problem* is where we want to know what value or values) of "x" that will solve the equation to yield a net zero value of the function.

Fixed Point Iteration

Fixed point iteration is one method to implement to tackle a *root solving problem*. These are problems where we find "zeros" of functions.

Fixed point iteration is one method of accomplishing root solving. To accomplish root solving by fixed point iteration, we must recast the equation in the form

$$x = g(x)$$

where we choose a starting value of $x = x_n$ so that we obtain x_{n+1}.

$$x_{n+1} = g(x_n)$$

If x_{n+1} is different from x_n by some small relative convergence tolerance ε, (such as 1E-4, or whatever is specified by the user), then the process is repeated. A typical criterion for convergence tolerance is

$$\varepsilon < \left| \frac{x_{n+1} - x_n}{x_n} \right|$$

Determination of Stability and Convergence for Fixed Point Iteration
Let $x = s$ be a solution of $x = g(x)$ on interval "J".

If $|g'(x)| \le k < 1$, then the iteration converges for any x_0 contained in "J".

Example

Determine the roots of $f(x) = x^3 + x - 1 = 0$ using iteration.
Applying fixed point iteration…

$$x(1+x^2) - 1 = 0$$
$$x(1+x^2) = 1$$
$$x = \frac{1}{1+x^2}$$
$$g(x) = \frac{1}{1+x^2}$$

Will this converge if $x_0 = 1$?

$$g'(x) = (1+x^2)^{-2}(2x) = \frac{-2x}{(1+x^2)^2}$$
$$x=1 \quad |g'(x)| = \frac{2(1)}{(1+1^2)^2} = \frac{2}{4} = 0.5 < 1$$

$\therefore$ fixed point iteration applied with this formulation should converge.

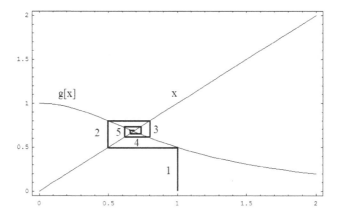

The above figure depicts a plot of x and $g(x)$, and the fixed point iteration marching sequence, effectively "spiraling in" to converge on the root, following the numerical iterative progression

The numerical iterative progression of $x_{n+1} = g(x_n)$ is given in the table below.

Table of Iterative Values for Fixed Point Iteration Example

Iterate	Value	Relative Difference
0	1.0	
1	0.5	0.5
2	0.8	0.6
3	0.609756	0.237805
4	0.728968	0.195507
5	0.653	0.104213
6	0.701061	0.0736013
7	0.670472	0.0436332
8	0.689878	0.0289436
9	0.677538	0.0178861
10	0.685374	0.0115642
11	0.680394	0.00726572
12	0.683557	0.00464898
13	0.681547	0.00294044
14	0.682824	0.00187354
15	0.682013	0.00118818
16	0.682528	0.000755775
17	0.682201	0.000479826
18	0.682409	0.000304997
19	0.682276	0.000193721
20	0.68236	0.000123102
21	0.682307	0.0000782031
22	0.682341	0.0000496896
23	0.682319	0.0000315685
24	0.682333	0.0000200575
25	0.682324	0.0000127432

The solution is 0.682324 for this marching example.

Bisection Method

The Bisection method is another method to obtain a solution to a root solving problem. In this method, for a given interval $[a, b]$, one tests for the case wherein $f(a) < 0$ and $f(b) > 0$, so that a and b "bracket" the root about zero, as guaranteed by the *intermediate value theorem*.

Then, one chooses

$$a_{n+1} = \frac{(a_n + b_n)}{2} \quad \text{or} \quad b_{n+1} = \frac{(a_n + b_n)}{2}$$

so that the interval continues to be "bisected" by values of opposite in sign, until $f(x) = 0$ is achieved by some tolerance criteria. This is graphically depicted in the figure below.

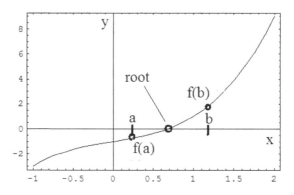

Graphical depiction of the Bisection Method

Newton's Method

Newton's Method is another way of root solving that is usually quite reliable, provided the initial guess is reasonable. We first try to compute $f(x_1)$ based on a local linearity estimate of $f(x_0)$, as follows:

$$f(x_1) \approx f(x_0) + \frac{df}{dx}\bigg|_{x \to x_0} (x_1 - x_0)$$

$$f(x_1) \approx f(x_0) + f'(x_0)(x_1 - x_0)$$

(Note that this is analogous to a truncated Taylor series, as presented in the chapter on Power Series). But, if $f(x_1) = 0$ because x_1 is assumed to be a root, then

$$0 \cong f(x_0) + f'(x_0)(x_1 - x_0) \quad \text{or} \quad -\frac{f(x_0)}{f'(x_0)} = x_1 - x_0$$

$$\text{and then} \quad x_1 = \left[x_0 - \frac{f(x_0)}{f'(x_0)} \right] \rightarrow g(x_n)$$

It should be noted that Newton's method may <u>diverge</u> if the initial guess is too far from the actual root.

Assuming the root is at a point where s is the root, converged as a result of assuming here that a small absolute tolerance applies:

$$\varepsilon = \left| f(s) - f(x_{n+1}) \right|,$$

From above, we find that if we take the derivative of $g(x)$:

$$g'(x) = \left[1 - \frac{f'(x)^2 - f(x)f''(x)}{f'(x)^2} \right] = \frac{f(x)f''(x)}{f'(x)^2}$$

We find that since $f(s) = 0$, then this yields $g'(s) = 0$
Then:

$$g''(s) = \frac{f''(s)}{f'(s)} \quad .$$

This relates to second order truncation in a Taylor's series, since Series,

$$\in_{n+1} \approx \frac{g''(\xi)}{2!}(s - x_n)^2 \approx \frac{f''(s)}{2f'(s)} \in_n^2$$

so that Newton's method yields *second order convergence* on the root.

Much more can be discussed on the order of convergence. For now, suffice it to say that higher order methods converge faster, and many methods, such as the Bisection method, are first order methods. This means that many more trials (iterations) will be necessary to yield a final result compared to higher order methods such as Newton's Method.

Numerical Integration Methods

Consider the fact that we wish to obtain the solution to an integration problem. Note this is practical when the functions pose a challenge to integrate. So, we would like to approximate the integral by using an *algebraic equation* rather than by doing a difficult integration; this is known as numerical "Quadrature," and so we seek a numerical solution to the problem of

$$J = \int_a^b f(x)dx$$

Where "J" is the area under the curve of function "f" between independent variables "a" and "b". (Note that here we assume that $f(x)$ is a "well-behaved" function, e.g. continuous, differentiable…, etc.)

Midpoint Rule

Approximate $f(x)$ using *n step functions* over *subintervals* that are "piecewise constant." In doing so, we have a constant step size

$$h = \frac{(b-a)}{n}$$

so that the integral is approximated by

$$J \approx f(x_1) \cdot h + f(x_2) \cdot h + \ldots + f(x_n) \cdot h$$

This is the *Rectangular or "Midpoint Rule"* (or "piecewise constant method") of numerical quadrature:

$$\text{Step Size} \left[h = \frac{(b-a)}{n} \right]$$

$$J_M \approx MID(n) = \sum_{i=1}^{n} f(x_i) \cdot h$$

Where values are taken to be at the center of the subinterval.

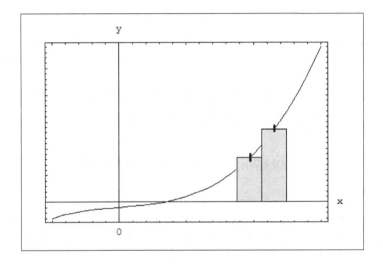

Graphical illustration of the midpoint quadrature method

Trapezoidal Rule

Once again, we want to solve the problem

$$J = \int_a^b f(x)dx$$

and consider a *constant step size* using

$$h = \frac{(b-a)}{n}$$

In this case, rather than using a constant for the function height at the center of the subinterval (as before), we now use an *average* of the function on the *left* and *right* sides of the subinterval, yielding a "Trapezoidal" shape.

$$J_T = TRAP(n) = \frac{1}{2}(f(a)+f(x_1))h + \frac{1}{2}(f(x_1)+f(x_2))h + ...\frac{1}{2}(f(x_{n-1})+f(b))h$$

Another way to write this is by using "Left" and "Right" to denote each side of the subinterval where the function is evaluated:

$$J_T = TRAP(n) = \frac{1}{2}(LEFT(n) + RIGHT(n))$$

This procedure is known as the *Trapezoidal Rule* (or "piecewise linear method"), and when the equation is simplified for n subintervals, then it can be equated to:

$$\text{Step Size} \left[h = \frac{(b-a)}{n} \right]$$

$$J_T = TRAP(n) = \sum_{i=1}^{n} h \cdot \left[\frac{1}{2}f(a) + f(x_1) + f(x_2) + ... + \frac{1}{2}f(b) \right]$$

It can be shown that an *error estimate* (ε) for the *Trapezoidal Rule* can be derived from a simple polynomial integration over a constant step size, and is given by:

$$\varepsilon \le \frac{(b-a)^3}{12n^2} f''(\tilde{x})$$

Up to now, we applied a piecewise constant over the subinterval, which led to the Rectangular Rule. Then applying a piecewise linear estimate led to the Trapezoidal Rule. Next, we will find that applying a piecewise quadratic will yield Simpson's Rule.

Applying a $p_2(x)$ LaGrange (interpolating) polynomial, we can perform *parabolic fits* to the *evenly spaced* subintervals that will *exactly solve* the integral up to order 3. Accomplishing this yields Simpson's Rule, which is equal to:

$$J_S = SIMP(n) = \frac{h}{3}\left[f(x_0) + 4f(x_1) + 2f(x_2) + 4f(x_3) + \ldots + 4f(x_{n-1}) + f(x_n)\right]$$

Example

Simpson's Rule with two subintervals on $[a, b]$.

$$\text{Step Size } \left[h = \frac{(b-a)}{2}\right]$$

$$SIMP(2) = \frac{h}{3}\left[f(a) + 4f(a+h) + f(b)\right]$$

An Important fact to recognize is that *Simpson's Rule* can directly be shown to be a weighted average of the *Midpoint* and *Trapezoidal Rules*:

$$J_S = SIMP(n) = \frac{1}{3}\left(2 \cdot MID(n) + 1 \cdot TRAP(n)\right)$$

Problem: Integrate $1/x$ from $[1,2]$ using 2 sub-intervals using each of the methods highlighted in this section.

Error in Quadrature Formulas

A thorough error analysis can be performed using Taylor's series with polynomial functions, which we will return to later. For now, the reader should consider that it can be shown that the error in approximating the integral for the Trapezoidal Rule has the opposite sign and twice the magnitude of the Midpoint Rule quadrature error. This is precisely why Simpson's rule uses twice the Midpoint Rule solution added to the Trapezoidal Rule solution, since the idea is to minimize the error through cancellation.

Also, in general, for Simpson's Rule, an extra 4 digits of accuracy requires ~10 times more work. For Midpoint and Trapezoidal Rule by themselves, each extra *two digits* of accuracy require ~10 times more work. Work" in this case refers to floating point operations, or "flops."

Gaussian Quadrature

Gaussian quadrature is used extensively in nuclear engineering in deterministic transport applications, since it is an extremely accurate method of approximating an integral with just a few evaluated points. Gauss, a talented mathematician, derived this quadrature method as a result of his search to yield high accuracy with the fewest evaluated points.

Again, we seek a numerical solution to the problem of

$$J = \int_a^b f(x)dx$$

Where "J" is the area under the curve of function "f" between independent variables "a" and "b" for a "well-behaved" function. Before we proceed further, we need to discuss two important items applied in Gaussian Quadrature: *Legendre Polynomials* and *Orthogonality*.

Legendre Polynomials

Legendre Polynomials are a family of polynomial functions that originate from the solution of Legendre's differential equation. In nuclear engineering, we encounter these polynomials (both Legendre Polynomials and *Associated* Legendre Polynomials) when considering neutron scattering theory in three-dimensional applications. Legendre's differential equation is:

$$(1-x^2)y''-2xy'+n(n+1)y = 0$$

Legendre functions have a unique property in that they are *orthogonal* on the interval $[-1,1]$; this is discussed at length in the next section.

Rodrigues' general formula for these polynomials is:

$$P_n(x) = \frac{1}{2^n\,n!}\frac{d^n}{dx^n}(x^2-1)^n$$

The first three Legendre Polynomials are given by:

$$P_0(x) = 1$$
$$P_1(x) = x$$
$$P_2(x) = \frac{1}{2}\left(3x^2 - 1\right)$$

Orthogonality of Functions

Orthogonality is a property we will use extensively in applications of this text. We have routinely encountered orthogonal coordinate systems, such as the Cartesian ("x, y, z") coordinates, which relate to linearly independent coordinates.

To discuss linearly independent functions, we need to discuss the concept of functional orthogonality. The formal definition of *orthogonality* of functions over a fixed interval $[a, b]$ with a weighting function $w(x)$ is:

$$\int_a^b \phi_i(x)\phi_j(x)w(x)dx = \delta_{ij}$$

where δ_{ij} is Kronecker Delta Function. The Kronecker Delta is a special function in mathematics defined to be:

$$\delta_{ij} = \begin{cases} 1 \text{ for } i = j \\ 0 \text{ for } i \neq j \end{cases}$$

The primary purpose of the weighting function $w(x)$ is to preserve ("guarantee") the orthogonality property on the interval.

A secondary purpose of the weighting function is to make orthogonal functions *orthonormal*, which makes the integration result so that it can be described by the Kronecker Delta function (1 or 0).

Now, as stated previously, the Legendre Polynomials are orthogonal on the interval $[-1,1]$, and they are *orthonormal* if used with the following weighting function for Legendre polynomials $P_n(x)$:

$$w(x) = \frac{(2n+1)}{2}$$

Other orthogonal functions include LaGuerre, Bessel's, and the Trigonometric functions.

The Gaussian Quadrature Theorem

Returning back to our discussion of quadratures, we would like to get an accurate evaluation of an integral using the *fewest number* of evaluated points on the interval.

So far, using the Midpoint, Trapezoidal, or Simpson's rule, we simply applied a uniform spacing (h) in the quadrature formulas over a number of intervals. So, what if we were able to choose specific values of x_i to evaluate $f(x_i)$, so that in effect, we choose just a few of the "right" locations that yielded high accuracy in approximating the integral?

This is what Gauss set out to do, and resulted in the Gaussian Quadrature Theorem. The *Gaussian Quadrature Theorem* states that if we have a polynomial of degree n, given by $\alpha_n(x)$ that has k real roots x_k (where the roots can be determined by root solving the polynomials using $\alpha_n(x_k) = 0$), and $\alpha_n(x)$ is *orthogonal* to a lower degree polynomial on the interval from $[a, b]$, so that

$$\int_a^b \alpha_n(x)\alpha_m(x)w(x)dx = \delta_{nm}$$

Then, it is possible to evaluate any integral on the interval $[a, b]$ at roots $x = \{x_k\}$ so that the following is true:

$$\int_a^b f(x)dx = \sum_{k=1}^n A_k f(x_k)$$

Also, from the equation above, that the integral expression will be <u>exact</u> for any real polynomial $f(x)$ of degree $(2n-1)$ or less.

Derivation of 2-Point Gaussian Quadrature

For this application, the polynomials we choose are Legendre Polynomials, orthogonal on $[-1,1]$. To evaluate two points on the interval, we need two roots, which implies we need the roots for $P_2(x)$, the 2nd degree Legendre polynomial.

$$\frac{1}{2}\left(3x^2 - 1\right) = 0$$

Therefore, roots are obtained by solving

$$3x^2 - 1 = 0$$

So we have:

$$x^2 = \frac{1}{3} \quad \text{with a result of} \quad x = \pm\frac{1}{\sqrt{3}}$$

So, considering $x_1 = \frac{1}{\sqrt{3}}$, and $x_2 = -\frac{1}{\sqrt{3}}$ (where $\frac{1}{\sqrt{3}} \approx 0.57735$):

$$A_1 x_1 + A_2 x_2 = \int_a^b x\,dx = \frac{b^2}{2} - \frac{a^2}{2}$$

$$A_1 x_1^2 + A_2 x_2^2 = \int_a^b x^2\,dx = \frac{b^3}{3} - \frac{a^3}{3}$$

Now we have 2 equations and 2 unknowns (A_1, A_2), and solving these equations simultaneously, we get:

$$A_1 = 1 \quad A_2 = 1$$

Then, for Two-point Gaussian Quadrature,

$$\int_{-1}^{1} x^2\,dx = 1\left(\frac{1}{\sqrt{3}}\right)^2 + 1\left(-\frac{1}{\sqrt{3}}\right)^2 = 2/3$$

The only problem now is that we will need to integrate over an interval that is not only over $[-1,1]$. To integrate over $[a, b]$ that is not $[-1,1]$, we need to scale $[a, b]$ onto $[-1,1]$, effectively projecting the interval.

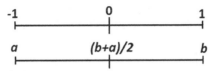

Figure depicting the projection of coordinates for Gaussian Quadrature

This is achieved by using the following formulation for 2-point Gaussian quadrature on an arbitrary interval $[a, b]$:

$$\frac{(b-a)}{2}\left[f\left(\frac{b+a}{2}+\frac{b-a}{2}x_1\right)+f\left(\frac{(b+a)}{2}+\frac{b-a}{2}x_2\right)\right]$$

Example

Compare the exact integral of $f(x) = x^2$ from $[3,7]$ with the result of 2-point Gaussian quadrature; the exact integral is:

$$\int_3^7 x^2 dx = \frac{x^3}{3}\Bigg]_3^7 = \frac{316}{3}$$

Using 2-point Gaussian Quadrature:

$$\int_3^7 x^2 dx = \left(\frac{7-3}{2}\right)\left[1\cdot f\left(5+\frac{2}{\sqrt{3}}\right)+1\cdot f\left(5+-\frac{2}{\sqrt{3}}\right)\right] = \frac{316}{3}$$

Gaussian quadrature can be derived for any number of points, and is also applicable in multiple dimensions.

NUCLEAR APPLICATION: Disk Radiation Source

The Solid Angle Ω for the geometry of a cylindrical detector detecting gamma rays emitted from a radioactive gamma disk source (Disk Source Radius R_s) at a distance d away from a circular radiation detector (Detector Radius R_d) is computed using the $J_1(x)$ Bessel function (derived later in this text) according to the formula:

$$\Omega = \lim_{\omega \to \infty} 4\pi \int_0^\omega R_d/R_s \exp(-\frac{\omega\ d}{R_s})\frac{J_1(\omega)}{\omega} J_1(\omega\ R_d/R_s)d\omega$$

$$\Omega_{fraction} = \lim_{\omega \to \infty} \int_0^\omega R_d/R_s \exp(-\frac{\omega\ d}{R_s})\frac{J_1(\omega)}{\omega} J_1(\omega\ R_d/R_s)d\omega$$

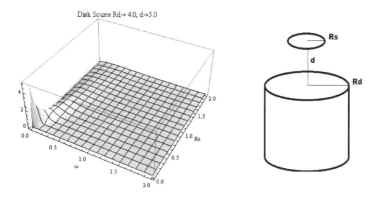

(Left) $\Omega_{fraction}$ plot, $R_d = 4.0$, $d = 5.0$ cm; (Right) Circular disk source, cylinder detector

Note: A reasonable numerical upper limit for ω is 15, and the Bessel Function can be accurately computed using the truncated power series:

$$J_1(x) = (x/2) - (x^3/16) + (x^5/384) - (x^7/18432) + (x^9/1474560) - (x^{11}/176947200) + (x^{13}/29727129600)$$

One can compute the solid angle by the following methods:

Write a FORTRAN or C language code using a subroutine for each of the following:

- Midpoint method 'MID' using the midpoint of each sub-interval.
- Trapezoidal method 'TRAP' using ½*(LEFT + RIGHT), referring to an average of the function evaluated on the "Left" and "Right" sides of each sub-interval.
- Simpsons Rule using the weighted mean of 1/3*(2*MID + TRAP).
- Each of the above tasks must be coded to use an <u>adaptive loop algorithm</u> that checks the accuracy of the calculation after first n sub-intervals, then by increasing the number of sub-intervals by some number (e.g. n + 20)), computing the relative change in the solution.
- Perform each calculation and report the number of sub-intervals required to achieve a relative error of 1.0E-06 according to

$$\varepsilon = \frac{\left| Integral_{n+5} - Integral_n \right|}{\left| Integral_n \right|}$$

There is an exercise for this problem in the appendix.

Many of the concepts discussed here will be applied throughout the remaining sections of the text. In the next chapter we briefly review fundamentals of complex numbers.

Chapter 4

Fundamentals of Complex Numbers

In order to understand several concepts used in the mathematics of solving problems in nuclear engineering, it is important to have experience with complex numbers. Complex numbers play a role in the solution of differential equations, as well as in spherical harmonics functions, and other applications involving nuclear engineering. Therefore, this chapter covers a very brief look at complex numbers.

Definition of Complex Numbers

Consider two real numbers x and y. These two real values are cast as a complex number in rectangular form composed of real and imaginary parts, where the number x is the real part multiplier, and the number y is the imaginary part multiplier:

$$z = x + iy \qquad i = \sqrt{-1} \rightarrow (0,1)$$

$$x = \text{Re}[z] \qquad y = \text{Im}[z]$$

Two complex numbers are equal when both real and imaginary multipliers are equal

$$z_1 = z_2 \quad IF \quad x_1 = x_2 \quad and \quad y_1 = y_2$$

Complex Number Operations

Addition is accomplished by

$$z_1 + z_2 \Rightarrow (x_1 + x_2) + i(y_1 + y_2)$$

Multiplication must be accomplished with close attention algebraically

$$z_1 \cdot z_2 \Rightarrow (x_1 + iy_1)(x_2 + iy_2)$$
$$z_1 \cdot z_2 = x_1 x_2 + y_1 y_2 (i)^2 + i(x_2 y_1 + x_1 y_2)$$
$$z_1 \cdot z_2 = (x_1 x_2 - y_1 y_2) + i(x_1 y_2 + x_2 y_1)$$

Complex numbers can be plotted on a 2-dimensional plane formed as an x-y axis system called the real and imaginary axes in Cartesian coordinates.

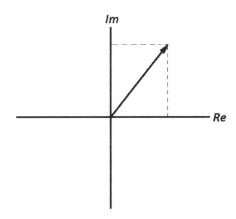

A complex number Z plotted on the complex plane.

The Complex Conjugate of $z = x + iy$ is denoted by $z^* = x - iy$ and is obtained by geometrically reflecting across the real axis.

Properties of $zz*$ $(z\bar{z})$

Note that $zz* = x^2 + y^2$, and that

$$z + z* = 2x \quad z - z* = 2iy$$

$$\mathrm{Re}\, z = x = \frac{1}{2}(z + z*) \quad \mathrm{Im}\, z = y = \frac{1}{2i}(z - z*)$$

Example

Consider the two complex numbers z_1 and z_2

$$z_1 = 4 + 3i$$
$$z_2 = 2 + 5i$$

Use complex conjugates to demonstrate the imaginary part of z1

$$\mathrm{Im}\, z_1 = \frac{1}{2i}[(4 + 3i) - (4 - 3i)] = \frac{3i + 3i}{2i} = 3$$

Note that $\mathrm{Im}(z_1) \rightarrow 3 \neq 3i$

Multiply the two numbers:

$$z_1 z_2 = (4 + 3i)(2 + 5i) = 8 + 6i + 20i - 15 = -7 + 26i$$

And by identity,

$$\bar{z}_1 \bar{z}_2 = (4 - 3i)(2 - 5i) = 8 - 6i - 20i - 15 = -7 - 26i$$

Polar form of complex numbers

Complex numbers can be expressed in polar form using the Euler formula. Recall that according to Euler,

$$z = x + iy \rightarrow r(\cos\theta + i\sin\theta)$$

And accordingly,

$$x = r\cos\theta$$

$$y = r\sin\theta$$

$$\therefore z = r(\cos\theta + i\sin\theta)$$

We note that $[r] \equiv$ the absolute value or "modulus" of "z", so that

$$|z| = r = \sqrt{x^2 + y^2} = \sqrt{z\bar{z}} = \sqrt{zz*}$$

And $|z|$ is the distance of the point in the complex plane to the origin.

$$[\theta] \equiv \text{argument of "z"} \Rightarrow \arg z = \arctan\frac{y}{x}$$

note that this angle result must be in radians. Also, Arg z falls between $-\pi < \theta \leq \pi$ and the capital "A" in "arg" denotes it is a "principle value".

Multiplication and Division in Polar form

Multiplication is accomplished in polar form as follows:

$$z_1 z_2 = r_1 r_2 [\cos(\theta_1 + \theta_2) + i\sin(\theta_1 + \theta_2)]$$

Which can be seen by following the steps indicated:

$$z_1 = r_1\left(\cos\theta_1 + i\sin\theta_1\right)$$

$$z_2 = r_2\left(\cos\theta_2 + i\sin\theta_2\right)$$

$$z_1 z_2 = r_1 r_2\left(\cos\theta_1 + i\sin\theta_1\right)\left(\cos\theta_2 + i\sin\theta_2\right)$$

$$z_1 z_2 = r_1 r_2\left(\cos\theta_1\cos\theta_2 - \sin\theta_1\sin\theta_2\right) + i\left(\sin\theta_1\cos\theta_2 + \cos\theta_1\sin\theta_2\right)$$

Applying sine, cosine addition rules, we obtain the desired result

$$z_1 z_2 = r_1 r_2\left(\cos\left(\theta_1 + \theta_2\right) + i\sin\left(\theta_1 + \theta_2\right)\right)$$

Complex Multiplication

$$\left[\left|z_1 z_2\right| = \left|z_1\right|\left|z_2\right|\right]$$

$$\left[\theta = \arg\left(z_1 z_2\right) = \arg z_1 - \arg z_2\right] \text{ up to multiples of } 2\pi$$

Complex Division

$$\left[\left|\frac{z_1}{z_2}\right| = \frac{\left|z_1\right|}{\left|z_2\right|}\right]$$

$$\theta = \arg\left(\frac{z_1}{z_2}\right) = \arg z_1 - \arg z_2$$

Polar form is important for several reasons. One of these is that computing powers of complex numbers, it is essential to simplify the operation:

Example

Given $z_3 = -1 - i$

Compute z_3^4...Well, we could use...$(-1-i)^4$...this is in fact quite cumbersome to perform using the rectangular form, but much easier in polar form, again employing Euler's formula, and finding the argument:

$$e^{i\theta} = \cos\theta + i\sin\theta \qquad z = re^{i\theta}$$

$$\theta = \arctan\frac{-1}{-1} = \frac{\pi}{4}$$

We note here that the original number falls into the third quadrant of the Complex Plane, so that we must add π radians to the argument

$$\therefore \theta = \frac{5\pi}{4}$$

Now, finding the modulus

$$r = |z_3| = \sqrt{1^2 + 1^2} = \sqrt{2}$$

$$z_3 = (-1-i) = \sqrt{2}\left(\cos\frac{5\pi}{4} + i\sin\frac{5\pi}{4}\right)$$

$$z_3 = \sqrt{2}e^{i5\pi/4}$$

$$z_3^4 = \left(\sqrt{2}e^{i5\pi/4}\right)^4 = 2^{4/2}e^{i5\pi} = 4e^{i5\pi}$$

And now we can immediately discern where this point is, at radius 4 and 5π rotations from the polar axis.

Complex Functions

Consider the complex function composed of real coefficients $u(x, y)$ and $v(x, y)$

$$w = f(z) = u(x, y) + iv(x, y)$$

Example

Given the complex function $w = f(z) = z^2 + 3z$

Find $u(x, y)$ and then compute $f(1 + 3i)$?

Recall:

$$u(x, y) = \text{Re}[f(z)] = \frac{1}{2}(f(z) + f(z)*)$$

But recall that the complex number

$$z = x + iy \qquad \text{Im}[f(z)] = \frac{1}{2i}(f(z) + f*(z))$$

then

$$f(z) = (x + iy)^2 + 3(x + iy)$$

And $u(x, y)$ is found from:

$$\frac{1}{2}\left((x + iy)^2 + 3(x + iy) + (x - iy)^2 + 3(x - iy)\right)$$

$$\frac{1}{2}\left(x^2 - y^2 + 2xyi + 3x + 3iy + x^2 - y^2 - 2xyi + 3x - 3iy\right)$$

$$\frac{(2x^2 - 2y^2 + 6x)}{2} = (x^2 - y^2 + 3x)$$

And we evaluate the function

$$f(1 + 3i) = (1 + 3i)^2 + 3(1 + 3i)$$
$$= 1 - 9 + 6i + 3 + 9i = [-5 + 15i]$$

Complex functions have limits, continuity, and therefore can have derivates. The limit must exist from both sides for the requirements for function continuity and differentiability. If the functions are not differentiable, then they are not considered to be analytic functions.

In general, simple real valued functions are *analytic* if a mixed partial derivative is fully consistent.

Example

Show $x^2 y$ is an analytic function

$$\frac{\partial^2 f}{\partial x \partial y} = \frac{\partial^2 f}{\partial y \partial x} = 2x \quad \text{since} \quad \frac{\partial}{\partial y}\frac{\partial f}{\partial x} = \frac{\partial}{\partial x}\frac{\partial f}{\partial y}$$

Now, in the case of Complex Numbers, if $f(z)$ is analytic, then $f(z)$ must satisfy the Cauchy–Riemann Equations

$$\text{Given} \quad f(z) = u(x, y) + iv(x, y)$$

$$\frac{\partial u}{\partial x} = \frac{\partial v}{\partial y} \qquad \frac{\partial u}{\partial y} = -\frac{\partial v}{\partial x}$$

In the domain D on {x,y}

Example

Show that z^2 is analytic

$$f(z) = z^2 \text{ so that } z^2 = (x + iy)(x + iy) = (x^2 - y^2) + 2ixy$$

$$u(x, y) = \text{Re}[z^2] = x^2 - y^2$$
$$v(x, y) = \text{Im}[z^2] = 2xy$$
and
$$\frac{\partial u}{\partial x} = 2x \qquad \frac{\partial v}{\partial y} = 2x$$
$$\frac{\partial u}{\partial y} = -2y \qquad -\frac{\partial v}{\partial x} = -2y$$

Thereby demonstrating the function is analytic. We can use Mathematica to demonstrate a similar result:

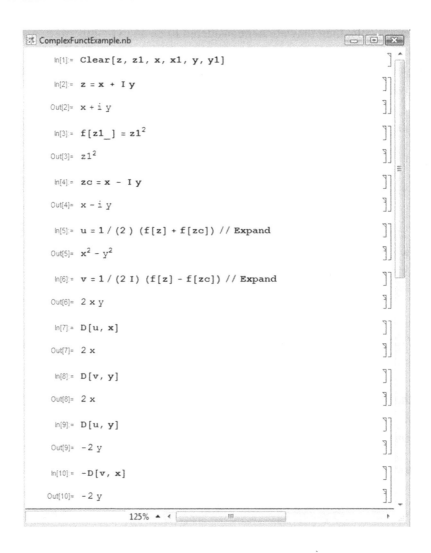

Example

Is $z = x - iy$ analytic?

To answer, we verify to determine if it satisfies the Cauchy–Riemann equations:

$$u = x \quad v = y$$

$$\frac{\partial u}{\partial x} = 1, \quad \frac{\partial v}{\partial y} = -1$$

$$\frac{\partial u}{\partial y} = 0, \quad \frac{\partial v}{\partial x} = 0$$

$$\therefore \text{no, } f(z) = z \text{ is not analytic}$$

Polar form of Cauchy–Riemann equations

With $z = r(\cos \theta + i \sin \theta)$ the Cauchy–Riemann Equations become:

$$\frac{\partial u}{\partial r} = \frac{1}{r}\frac{\partial v}{\partial \theta} \qquad \frac{\partial v}{\partial r} = -\frac{1}{r}\frac{\partial u}{\partial \theta}$$

Complex exponential functions

Example

Given z is a complex number $= x + iy$

$$e^z = e^x(\cos y + i \sin y) \qquad \left[e^{z_1 + z_2} = e^{z_1} e^{z_2} \right]$$

$$\text{if } z_1 = x \quad z_2 = iy$$

$$e^{z_1 + z_2} = e^x e^{iy} = e^x(\cos y + i \sin y)$$

Then we have $z = re^{i\theta}$

Note: Evaluating $e^{2\pi i} = (\cos 2\pi + i \sin 2\pi) = 1$

$$e^{\pi i} = (\cos \pi + i \sin \pi) = -1$$

Chapter 5

Methods for Solving Ordinary Differential Equations

A differential equation, according to the *Dictionary of Mathematics*, is:

> An Ordinary Differential Equation (ODE) is an equation that defines a relationship between an independent variable x and a dependent variable y, and one or more derivatives of y with respect to x, the solution of which leads to a mathematical identity when the solution is substituted back into the equation.

Also, a Partial Differential Equation (PDE) is a differential equation that involves more than one independent variable, with dependent variable partial derivatives of the independent variables in an equation.

In engineering and science, most processes can be described or modeled using differential equations, and engineers must be familiar with all aspects of how to yield a solution to all types of differential equations.
In nuclear engineering, all aspects of reactor physics involve both ordinary and partial differential equations:

- Production, decay, and mixing of isotopes
- Burnup inventory of nuclear fuel
- Neutron diffusion, transport
- Criticality safety and reactivity management
- Heat transfer and thermal hydraulics

This chapter begins with some mathematical preliminaries that must be understood prior to launching into differential equations solution methods. Then we will work a number of solution approaches and examples.

The Dirac Delta Function

The Dirac Delta function is a tool often used to define unique functional behavior, such as a point source in space, or similar relationships.

Consider

$$f(r)\delta(r - r_0)$$

Where

$$\delta(r - r_0)$$

is a Dirac Delta function

Recall the *Kronecker* Delta function was used in the definition of orthogonality:

$$\int_a^B \phi_i(x)\phi_j(x)w(x)dx = \delta_{ij} = \begin{cases} 0 & i \neq j \\ 1 & i = j \end{cases}$$

The Dirac Delta function is quite different, and the reader should understand if the delta function is a Dirac delta function by the manner in which the equation is used.

For the Dirac Delta function, $\delta(r - r_0)$ is undefined (infinite) at $r = r_0$, and is 0 elsewhere. The *integral* of a Dirac Delta function is *finite*, provided $f(r_0)$ exists, and is defined by

$$\int_a^b f(r)\delta(r - r_0)dr = \begin{cases} f(r_0) & a < r_0 < b \\ \dfrac{1}{2}f(r_0) & a = r_0 \\ 0 & otherwise \end{cases}$$

Example

Given a point source in space, the source density $s(r)$ of neutrons in units of $\dfrac{n}{cm^3 s}$ is

$$s(r) = \frac{s_0 \delta(r)}{2\pi r^2}$$

Where in light of the definition for the Dirac Delta function, the units must be as follows:

s_0 n/s for the point source located at the origin

$\delta(r)$ 1/cm   for the Dirac Delta function

since if we integrate the source density $s(r)$ over the volume, then the number of neutrons emitted from the source term must be, by conservation:

$$\int_V s(r)dV \equiv s_0$$

Where $dV = r^2 \sin\theta \, d\theta \, d\phi \, dr$

Performing the integration of the source density to obtain the total source strength:

$$\int_0^R \int_0^{2\pi} \int_0^{\pi} \frac{s_0 \delta(r)}{2\pi r^2} r^2 \sin\theta \, d\theta \, d\phi \, dr$$

or

$$\int_0^R \frac{s_0 \delta(r)}{2\pi r^2} 4\pi r^2 = \int_0^R 2s_0 \delta(r) dr = 2s_0 \left(\frac{1}{2}\right) = s_0$$

The Dirac Delta function provides a mathematical vehicle by which we can define real world values in a consistent formulation to solve problems.

Classification of ODEs

Ordinary differential equations (ODE)s are classified by the order of the highest order derivative in the equation. The degree of the equation refers to the highest power of the highest order derivative.

Example

Consider the following second order ordinary differential equation of the first degree:

$$y'' + 3y' + y = 0$$

Linear vs. Non-linear ODEs

The following ODEs are "linear" and "non-linear", respectively:

$$\text{linear: } x^3 \frac{d^3 y}{dx^3} + 5y = e^x \qquad \text{non-linear: } \frac{d^3 y}{dx^3} + y^2 = 0$$

The second equation is non-linear due to the y^2 term in the equation.

Example

The following equation is of order 2, degree 3/2, non-linear, and homogeneous:

$$(y'')^{3/2} + (y')^3 + y = 0$$

ODE Standard Form

In the standard form for a linear ODE, the dependent variable "y" and all its derivatives are of the first degree, and each coefficient of each variable depends only on "x". This can be expressed by the following equation:

$$a_n(x)\frac{d^n y}{dx^n} + a_{n-1}(x)\frac{d^{n-1} y}{dx^{n-1}} + ... + a_1(x)\frac{dy}{dx} + a_0(x)y = g(x)$$

Separable ODEs

These ODEs contain variables that can be readily separated (e.g. "x" and "y") on opposite sides of the "$=$" sign, and subsequently solved by integrating both sides directly.

Example

The time rate of decay of radioactive atoms in a sample is proportional to the amount present. Expressed mathematically, this is a homogeneous ODE that is separable:

$$\frac{dN}{dt} + N\lambda = 0$$

Where N radioactive atoms exist at any time t, *and* $\lambda \equiv$ decay probability per unit time. If we assume that the amount at time $t = 0$ is $N(0) = N_0$, then we have an initial value problem. This is solved by separating the variables for subsequent integration:

$$\frac{dN}{dt} = -N\lambda$$

$$\int \frac{dN}{N} = \int -\lambda dt$$

$$\ln N = -\lambda t + c_1$$
$$N(t) = e^{-\lambda t + c_1} = e^{c_1}e^{-\lambda t} = ce^{-\lambda t}$$
$$N(t) = ce^{-\lambda t}$$
$$N(0) = N_0$$
$$N_0 = ce^0 \quad c = N_0$$
$$\text{then} \quad \left[N(t) = N_0 e^{-\lambda t} \right]$$

It is customary that half-life, the time it takes for half of the sample to decay, is given. One can solve for the decay constant from the half life by:

$$\frac{1}{2} = \frac{N(t)}{N_0} = e^{-\lambda T_{\frac{1}{2}}}$$

$$\ln \frac{1}{2} = -\lambda T_{\frac{1}{2}}$$

$$\ln 2^{-1} = -\lambda T_{\frac{1}{2}}$$

$$-\ln 2 = -\lambda T_{\frac{1}{2}}$$

$$\boxed{\lambda = \frac{\ln 2}{T_{\frac{1}{2}}}}$$

Example

$$Am^{242m} \left(T_{\frac{1}{2}} = 152\, y\right)$$

is a fissile isotope that is ten times more reactive than

$$Pu^{239} \left(T_{\frac{1}{2}} = 24,110\, y\right)$$

considering the fission cross sections of both of these isotopes. What fractions of Am^{242m} and Pu^{239} remain after 20 y?

Pu: $\dfrac{N(20)}{N_0} = e^{-(\ln(2)/24110)(20)} = 0.999$

Am: $\dfrac{N(20)}{N_0} = e^{-(\ln(2)/152)(20)} = 0.913$

More on radioactive decay will be discussed in other examples later in the chapter.

"Exact Differential" Equations

Consider the real valued equation

$$M(x, y)dx + N(x, y)dy = 0$$

This implies an "exact" differential equation. If we assume $u(x, y)$ then the total differential of $u(x, y)$, from the chain rule, is

$$du = \frac{\partial u}{\partial x} dx + \frac{\partial u}{\partial y} dy$$

Then if u $u(x, y)$ is "well behaved" over the domain, then mixed partial derivatives can be utilized:

$$\underbrace{\frac{\partial}{\partial y}\left(\frac{\partial u}{\partial x}\right)}_{M(x, y)} = \underbrace{\frac{\partial}{\partial x}\left(\frac{\partial u}{\partial y}\right)}_{N(x, y)}$$

So that

$$\frac{\partial^2 u}{\partial y \partial x} = \frac{\partial^2 u}{\partial x \partial y}$$

Then a "Test" for exactness is applied given the original equation; We determine if

$$\frac{\partial M}{\partial y} = \frac{\partial N}{\partial x}$$

If this is true, then the original differential equation must have been derived from an exact differential...

$$\frac{\partial f}{\partial x} = M \qquad \frac{\partial f}{\partial y} = N$$

Example

Solve for $f(x,y)$ if

$$\left(6xy^3 + \cos y\right)dx + \left(9x^2 y^2 - x\sin y\right)dy = 0$$

Check for Exactness:

$$\frac{\partial M}{\partial y} = 18xy^2 + -\sin y \qquad \frac{\partial N}{\partial x} = 18xy^2 - \sin y$$

We conclude this is an exact differential equation, and then assume that we can write

$$\frac{\partial f}{\partial x} = 6xy^3 + \cos y \qquad\qquad \frac{\partial f}{\partial y} = 9x^2 y^2 - x\sin y$$

We then have

$$f(x, y) = \int \left(6xy^3 + \cos y\right)dx \qquad f(x, y) = \int \left(9x^2 y^2 - x\sin y\right)dy$$

$$3x^2 y^3 + x\cos y + g(y) \qquad\qquad 9x^2\frac{y^3}{3} + x\cos y + h(x)$$

$$\left[f(x, y) = 3x^2 y^3 + x\cos y + C\right]$$

Simplifications by Substitution

Sometimes we can simplify a differential equation by substitution, e.g. using some $u = f(x,y)$ but always maintaining the independent variable x.

$$x\frac{dy}{dx} - y = \frac{x^3}{y}e^{y/x}$$

where x is the independent variable

$$\frac{dy}{dx} = \frac{y}{x} + \frac{x^2}{y}e^{y/x}$$

Try $u = \dfrac{y}{x}$; then we get, as an intermediate step, $\dfrac{dy}{dx} = u + \dfrac{x}{u}e^u$

This can be further simplified as follows:

$$y = ux$$

$$\frac{dy}{dx} = u'x + u$$

$$u'x + u = u + \frac{x}{u}e^u$$

$$u'x = \frac{x}{u}e^u$$

$$\frac{du}{dx} = u' = \frac{e^u}{u}$$

So that the equation that must be solved is

$$\frac{du}{dx} = \frac{e^u}{u}$$

which simplifies to $\int ue^{-u}du = \int dx$

Recall the formulation for integration by parts:

$$\int u_1 dv_1 = u_1 v_1 - \int v_1 du_1$$

With this notation, it means that

$$\begin{aligned} u_1 &= u \\ dv_1 &= e^{-u}du \end{aligned} \quad \text{which leads to} \quad \begin{aligned} du_1 &= du \\ v_1 &= -e^{-u} \end{aligned}$$

Substituting "by parts" and integrating both sides of the equation yields

$$-ue^{-u} + \int e^{-u} du = x + c$$

$$ue^{-u} + e^{-u} = -x + c$$

$$e^{-u}(u+1) = -x + c$$

$$u + 1 = (c - x)e^{u}$$

Now we back-substitute u:

$$\frac{y}{x} + 1 = c - xe^{y/x}$$

The final result is an implicit equation:

$$\left[y + x = x(c_1 - x)e^{y/x} \right]$$

Integrating Factor—Growth and Decay Problems

Often an integrating factor can be successfully applied to solve an ODE by creating an exact derivative to permit direct integration to determine a solution.

$$\text{Given} \quad f(x)\frac{dy}{dx} + g(x)y = h(x)$$

First, we need to make the coefficient of $\frac{dy}{dx}$ unity.

This yields $\dfrac{dy}{dx} + \dfrac{g(x)}{f(x)} y = \dfrac{h(x)}{f(x)}$

which in standard form is

$$\frac{dy}{dx} + P(x)y = Q(x)$$

The integrating factor is obtained by computing

$$I = e^{\int P(x)dx}$$

We multiply each term in the equation by the integrating factor I. The left hand side (LHS) can then be cast as an exact derivative, permitting direct integration of both sides.

Example

$$\text{Solve the ODE:} \quad \frac{dy}{dx} + 2xy = x \quad y(0) = -3$$

We note the integrating factor, and apply it to the ODE:

$$P(x) = 2x$$

$$e^{\int P(x)dx} = e^{\int 2xdx} = e^{2x^2/2} = e^{x^2}$$

$$e^{x^2}\frac{dy}{dx} + 2xye^{x^2} = xe^{x^2}$$

This can be recast as an exact differential

$$\frac{d}{dx}\left(e^{x^2}y\right) = xe^{x^2}$$

Solving,

$$\int d\left(e^{x^2}y\right) = \int xe^{x^2}dx$$

$$\rightarrow u = x^2 \quad du = 2xdx \quad \frac{du}{2} = xdx$$

We then obtain

$$e^{x^2} y = \frac{1}{2} e^{x^2} + c \quad \text{or} \quad y = \frac{1}{2} + c e^{-x^2}$$

Finally, we apply the initial condition:

$$y(0) = -3$$

$$-3 = \frac{1}{2} + c$$

$$c = -\frac{7}{2}$$

So that the complete solution is

$$y = \left[\frac{1}{2} - \frac{7}{2} e^{-x^2} \right]$$

NUCLEAR APPLICATION: Coupled Growth and Decay

Multiple linked cases of decay and growth offer an excellent Example of Integrating Factor. Consider the decay of lead-211, which leads to bismuth, thallium, and eventually, stable lead, as shown in the decay sequence below:

$$^{211}_{82}Pb \underset{\substack{36.1 \\ \text{min}}}{\overset{\beta^-}{\rightarrow}} {}^{211}_{83}Bi \underset{\substack{2.15 \\ \text{min}}}{\overset{\alpha}{\rightarrow}} {}^{207}_{81}Tl \underset{\substack{4.79 \\ \text{min}}}{\overset{\beta^-}{\rightarrow}} {}^{207}_{82}Pb \; _{Stable}$$

Assumptions: At time zero, we "freshly purify" ^{211}Pb, and no other sources of ^{211}Pb are present.

The differential equations describing the process of the time rate of change of the various isotopes are:

$$\frac{dN_1}{dt} = -\lambda_1 N_1 \qquad\qquad N_1(0) = N_1^0$$

$$\frac{dN_2}{dt} = \lambda_1 N_1 - \lambda_2 N_2 \qquad N_2(0) = 0$$

$$\frac{dN_3}{dt} = \lambda_2 N_2 - \lambda_3 N_3 \qquad N_3(0) = 0$$

$$\frac{dN_4}{dt} = \lambda_3 N_3 \qquad\qquad N_4(0) = 0$$

The first equation was already solved as an example of a separable equation to yield a solution as

$$\left[N_1(t) = N_1^0 e^{-\lambda_1 t} \right]$$

Now consider growth and decay of ^{211}Bi

$$\frac{dN_2}{dt} = \lambda_1 N_1 - \lambda_2 N_2$$

or rewriting the equation as

$$\frac{dN_2}{dt} + \lambda_2 N_2 = \lambda_1 N_1$$

Clearly, this is an integrating factor problem, where this factor is

$$e^{\int \lambda_2 dt} \rightarrow e^{\lambda_2 t}$$

where we multiply through the equation

$$e^{\lambda_2 t}\left(\frac{dN_2}{dt} + \lambda_2 N_2 = \lambda_1 N_1\right)$$

$$\frac{dN_2}{dt}e^{\lambda_2 t} + \lambda_2 N_2 e^{\lambda_2 t} = \lambda_1 N_1^0 e^{-\lambda_1 t} e^{\lambda_2 t}$$

Creating an exact differential on the left side, we then integrate both sides to obtain a solution

$$\int \frac{d}{dt}\left(N_2 e^{\lambda_2 t}\right) dt = \int \lambda_1 N_1^0 e^{(\lambda_2 - \lambda_1)t} dt$$

$$N_2 e^{\lambda_2 t} = \frac{\lambda_1 N_1^0}{(\lambda_2 - \lambda_1)}\left(e^{(\lambda_2 - \lambda_1)t}\right) + \tilde{c}_2$$

$$N_2(t) = \frac{\lambda_1 N_1^0}{(\lambda_2 - \lambda_1)}e^{(\lambda_2 - \lambda_1)t}e^{-\lambda_2 t} + \tilde{c}_2 e^{-\lambda_2 t}$$

$$N_2(t) = \frac{\lambda_1 N_1^0}{(\lambda_2 - \lambda_1)}e^{-\lambda_1 t} + c_2 e^{-\lambda_2 t}$$

Next, we recall that the initial amount of ^{211}Bi at the start is zero:

$$N_2(0) = 0 = \frac{\lambda_1 N_1^0}{(\lambda_2 - \lambda_1)}(1) + c_2$$

$$\therefore c_2 = \frac{-\lambda_1 N_1^0}{(\lambda_2 - \lambda_1)}$$

Therefore, we solve for the solution for ^{211}Bi :

$$N_2(t) = \frac{\lambda_1 N_1^0}{(\lambda_2 - \lambda_1)} e^{-\lambda_1 t} + \frac{-\lambda_1 N_1^0}{(\lambda_2 - \lambda_1)} e^{-\lambda_2 t}$$

or

$$\left[N_2(t) = \frac{\lambda_1 N_1^0}{(\lambda_2 - \lambda_1)} \left(e^{-\lambda_1 t} - e^{-\lambda_2 t} \right) \right]$$

As an exercise, it can be shown that for ^{207}Tl

$$N_3(t) = \lambda_1 \lambda_2 N_1^0 \left[\frac{e^{-\lambda_1 t}}{(\lambda_2 - \lambda_1)(\lambda_3 - \lambda_1)} + \frac{e^{-\lambda_2 t}}{(\lambda_1 - \lambda_2)(\lambda_3 - \lambda_2)} + \frac{e^{-\lambda_3 t}}{(\lambda_1 - \lambda_3)(\lambda_2 - \lambda_3)} \right]$$

In addition, N_4 is stable, therefore it can come from a material balance (conservation of mass) determined as follows:

$$\left[N_4(t) = N_1^0 - \left(N_1^0 e^{-\lambda t} \right) - N_2(t) - N_3(t) \right]$$

One can see the recurring pattern for this series of decays; for this reason, *Bateman* derived the following for a starting amount of a nuclide $(N_1(t))$ when no other atoms are initially present.

$$\left[\begin{array}{l} (for\ i > 1) \\[2mm] N_i = N_1^0 \lambda_1 \lambda_2 ... \lambda_{i-1} \sum_{j=1}^{i} \frac{e^{-\lambda_j t}}{\prod\limits_{\substack{k=1 \\ k \neq j}}^{i} (\lambda_k - \lambda_j)} \end{array} \right]$$

If the are other atoms in the chain initially present, denoted by N_l^0, then the solutions to the Bateman chain equations become

$$
N_i = \sum_{l=1}^{i-1} \left[N_l^0 \lambda_l \lambda_{l+1} ... \lambda_{i-1} \sum_{j=l}^{i} \frac{e^{-\lambda_j t}}{\prod_{\substack{k=l \\ k \neq j}}^{i} (\lambda_k - \lambda_j)} \right] + N_i^0 e^{-\lambda_i t}
$$

NUCLEAR APPLICATION: *Criticality Safety Solution Mixing*

A critical mass is one where fissile fuel material (U-235, U-233, or Pu-239) is assembled in a volume where the multiplication of neutrons occurs in a chain reaction from fission in the mass, and this production of neutrons is precisely balanced by leakage out of the system and absorption in the mass; this will be discussed further when we consider neutron balance using both radiation transport and neutron diffusion.

Therefore, "criticality" is achieved when a critical mass of fuel is assembled; if more than enough fuel is available to where the neutron production over-runs leakage and absorption rates, this is a supercritical condition.

If not enough material is present to limit neutron production from fission compared to leakage and absorption rates in the mass, this is a subcritical conditions.

In reality, we only want to achieve criticality in a nuclear reactor, since assembling a critical mass in an unprotected environment can release a lethal dose of neutron and gamma radiation from the leaking neutrons and gamma radiation that accompanies such an event (for example, see the *Louis Slotin accident* in the literature).

Since the fissile fuels are often handled in solutions, usually nitrates after dissolution in nitric acid, where the presence of water and hydrogen atoms can enhance their probabilities for absorption by causing the neutrons to lose energy (resulting in a higher neutron cross section), mixing problems often occur so that one must prevent criticality by limiting the concentration of the fuel in a given scenario.

Example

Consider a certain 4-liter tank holds 300 cc of 4 g/cc plutonium nitrate solution. A criticality will occur if the tank reaches 10 g/cc. The tank is continually stirred with a propeller mixer to keep the solution in the tank well mixed.

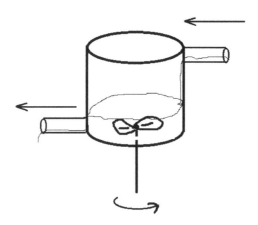

Graphic of 4 liter tank containing initial 300 cc of 4g/cc mixed plutonium bearing liquid

At time $t = 0$, a worker turns on the flow to add 12 g/cc of Pu nitrate solution at a rate of 2.5 cc/min. At that same moment, 1.0 cc/min of well mixed solution is withdrawn from the tank to flow out to another process stream.

Problem: Solve the differential equation for mixing to determine what happens in the tank as time passes. (Plot the concentration as a function of time in the tank using your solution). If the shift supervisor states that the mixing operation should terminate 150 minutes after it starts for criticality safety, and asks that for additional safety protocols, you check his/her conclusions. Do you concur?

To solve this problem, we need to determine the amount of mass of plutonium in the tank at any given time. The flow conditions are described by the following assembly of terms in a differential equation, where $Q[t]$ is the mass in grams of the plutonium in the tank at any time t based on inflow, outflow, and dilution relative to a constant flowrate in and out of the tank:

$$Q'[t] = \frac{12.g}{cc}\frac{2.5cc}{min} - \frac{Q[t]g}{(300cc + (2.5t)cc - (1.t)cc)}\frac{1cc}{min}$$

With the time zero amount $Q[0] = 300 * 4 = 1200$ g

The differential equation simplifies to

$$Q'[t] = 30. - \frac{Q[t]}{300 + 1.5t}$$

Clearly, this is an integrating factor problem, and

$$Exp[\int \frac{1}{300 + 1.5t} dt]$$

Which simplifies to

$$(300 + 1.5t)^{2/3}$$

Using this to solve the equation results in

$$Q[t] = -\frac{82078.85}{(200+t)^{2/3}} + 18.(200+t)$$

To obtain a concentration, we again divide by the volume

$$Qconc[t] = Q[t]/(300+1.5t) \text{ in g/cc}$$

Which results in the following concentration plot

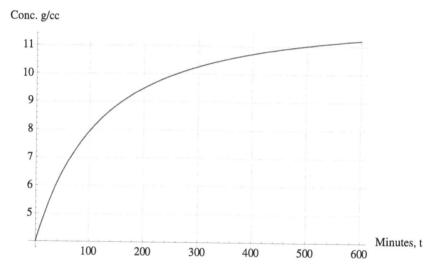

Conc. g/cc

Minutes, t

From graph, we cross the criticality point of 10g/cc at ~260 min, so one should concur with the shift supervisor that stopping the operation at 150 minutes is safe for preventing a criticality.

Root solving $Qconc[t] = 10\,g/cc$ yields a value of 259.5 minutes, just to confirm the above conclusions. These types of mixing problems are common in nuclear fuel and chemical processing flowsheets, where it is extremely important to understand process concentrations for mass limits, material safety limits, and criticality limits.

Higher order ODEs

An nth order linear homogeneous differential equation $L(y) = 0$ has "n" linearly independent solutions.

Linear independence

A set of functions is linearly independent on an interval $a \le x \le b$ if the only constants that satisfy

$$c_1 y_1(x) + c_2 y_2(x) + c_3 y_3(x) + \ldots = 0$$

are $c_1 = c_2 = \ldots = c_n = 0$, so that no linear combination of constants can be made to equate zero.

Example

Is the set $\{1, x, x^2\}$ independent on $(-\infty, \infty)$?

Consider the equation

$$c_1 + c_2 x + c_3 x^2 = 0$$

This is valid for all values of x only if

$$c_1 = c_2 = c_3 = 0$$

Therefore, the set of equations is linearly independent.

The Wronskian

The Wronskian $\left(W\left(y_n\right)\right)$ can be used to verify linear independence on an interval $a \le x \le b$ for the solutions y_n of $L(y) = 0$, a homogeneous differential equation.

The Wronskian is the determinant of increasing derivatives of the function being tested:

$$\text{For } W(y_n) \neq 0 \qquad W(y_n) = \begin{vmatrix} y_1 & y_2 & y_3 & \cdots \\ y_1' & y_2' & y_3' & \cdots \\ y_1'' & y_2'' & y_3'' & \cdots \end{vmatrix}$$

$$\vdots \quad \vdots \quad \vdots$$

If the Wronskian is <u>non-zero</u> for y_n's over the interval $a \leq x \leq b$, where y_n's must be "well-behaved" (continuous, differentiable, and finite over the interval).

Example

Use the Wronskian to verify the linear independence of $\sin(x)$, $\cos(x)$ on $-\infty < x < \infty$

$$\begin{vmatrix} \sin x & \cos x \\ \cos x & -\sin x \end{vmatrix} \Rightarrow -\sin^2 x - \cos^2 x = -\left(\sin^2 x + \cos^2 x\right) = -1 \neq 0$$

Therefore, the two solutions are linearly independent

Note: the Wronskian does *not* apply to an *arbitrary set of functions* – it applies *only* to those functions that are a solution to the homogeneous ODE $L(y) = 0$. For arbitrary functions we must use the definition of *linear independence*.

Example

Solve the equation (a homogeneous ODE):

$$y'' - y = 0$$

and demonstrate that the functions (solutions) are linearly independent…

$$y'' - y = 0$$
$$Try \ y = e^{mx}$$
$$y' = me^{mx} \quad y'' = m^2 e^{mx}$$

$$m^2 e^{mx} - e^{mx} = 0$$

$$e^{mx}\left(m^2 - 1\right) = 0$$
$$m^2 = 1 \quad m = \pm 1$$
$$So, \quad y_1 = e^x \quad y_2 = e^{-x}$$
$$y(x) = c_1 e^x + c_2 e^{-x}$$

And this is the general solution on $-\infty < x < \infty$. Now compute the Wronskian

$$W = \begin{vmatrix} e^x & e^{-x} \\ e^x & -e^{-x} \end{vmatrix} = -\left(e^x e^{-x}\right) - \left(e^x e^{-x}\right)$$
$$= -1 - 1 = -2 \neq 0$$

We can therefore conclude that the solutions are linearly independent.

The Hyperbolic functions: sinh(x) and cosh(x)

Returning to the previous ODE Example

$$y'' - y = 0 \quad \text{and if we try} \quad y = e^{mx}$$

Then

$$m^2 - 1 = 0$$
$$m = \pm 1 \quad \rightarrow \quad y_1 = c_1 e^{-x} + c_2 e^x$$

This same solution can be rewritten using hyperbolic functions

$$y = \alpha_1 \cosh(x) + \alpha_2 \sinh(x)$$

$$\alpha_1 = c_1 + c_2 \qquad \alpha_2 = c_1 - c_2$$

$$Because \quad \cosh(m_1 x) = \frac{1}{2}\left(e^{m_1 x} + e^{-m_1 x}\right)$$

$$\sinh(m_1 x) = \frac{1}{2}\left(e^{m_1 x} - e^{-m_1 x}\right)$$

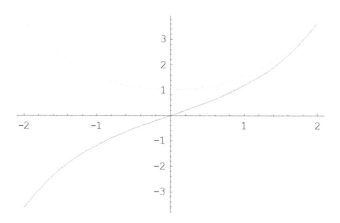

Plot of Sinh[x] (odd) and Cosh[x] (even) vs x

The hyperbolic functions sinh (mx) and cosh(mx) have useful properties, where $\cosh(0) = 1$ and $\sinh(0) = 0$, analogous to the values of cosine and sine functions, respectively.

The solution to the ODE expressed using hyperbolic functions is typically much more straight forward when solving the equation for the case of *finite* boundary conditions. For the case where one boundary may be infinite, the exponential form rather than the hyperbolic form should be utilized.

Different Solution Basis

Example

Compare $\sinh(x), \cosh(x)$ with e^{-x}, e^{x}

$$y'' - y = 0 \qquad y(0) = 1$$
$$y(1) = 2$$
$$m^2 - 1 = 0$$
$$m = \pm 1$$

As long as you have a set of equations that satisfy the differential equation over the interval of interest (linearly independent, analytic) these form a basis for solution of the differential equation. Consider both solutions to the differential equation:

$$\left[y = c_1 e^x + c_2 e^{-x} \right]$$

or

$$\left[y = c_3 \cosh(x) + c_4 \sinh(x) \right]$$

$$\cosh x = \left(\frac{e^x + e^{-x}}{2} \right)$$

$$\sinh x = \left(\frac{e^x - e^{-x}}{2} \right)$$

$$y(0) = 1 = c_1 + c_2 \qquad c_2 = 1 - c_1$$
$$y(1) = c_1 e^1 + c_2 e^{-1} \qquad c_1 e^1 + (1 - c_1) e^{-1} = 2$$
$$c_1 \left(e^1 - e^{-1} \right) = 2 - e^{-1}$$
$$\left[c_1 = \left(\frac{2 - e^{-1}}{\left(e^1 - e^{-1} \right)} \right) \right] \quad \left[c_2 = \left(1 - \left(\frac{2 - e^{-1}}{e^1 - e^{-1}} \right) \right) \right]$$

$$c_3(1) + c_4(0) = 1$$
$$\therefore c_3 = 1$$
$$\cosh(1) + c_4 \sinh(1) = 2$$
$$\therefore c_4 = \frac{2 - \cosh(1)}{\sinh(1)}$$

Standard Power Series Method

A straight forward way to solve any differential equation is via a power series approach. To characterize this method, we present the following example:

$$\text{Solve} \quad y'' + y = 0$$

Where we note this equation is valid for all real numbers.

To solve this equation following the traditional approach:

$$y = e^{mx}$$
$$y' = me^{mx} \quad y'' = m^2 e^{mx}$$

$$m^2 e^{mx} + e^{mx} = 0$$
$$m^2 = -1$$

and with application of Euler's formula, we obtain the general solution

$$y = c_1 \cos(x) + c_2 \sin(x).$$

Now consider the solution we would obtain using the power series method of solution at $x_0 = 0$. We begin by assuming the power series form, and taking derivatives:

$$y = a_0 + a_1 x + a_2 x^2 + a_3 x^3 + a_4 x^4 + \ldots$$
$$y' = a_1 + 2a_2 x + 3a_3 x^2 + 4a_4 x^3 + \ldots$$
$$y'' = 2a_2 + 6a_3 x + 12a_4 x^2 + \ldots$$

Then we substitute the series forms into the equation

$$\left(2a_2 + 6a_3 x + 12a_4 x^2 + \ldots\right) + \left(a_0 + a_1 x + a_2 x^2 + \ldots\right) = 0$$
$$\left(2a_2 + a_0\right) + \left(6a_3 + a_1\right)x + \left(12a_4 + a_2\right)x^2 + \left((n+2)(n+1)a_{n+2} + a_n\right)x^n = 0$$

$$a_n = -(n+2)(n+1)a_{n+2}$$

$$a_{n+2} = \frac{-a_n}{(n+2)(n+1)}$$

$$n = 0 \rightarrow n = 2 \rightarrow n = 4 \qquad n = 1 \rightarrow n = 3 \rightarrow n = 5$$

We obtain a general solution and assume

$$a_0 = 1$$
$$a_1 = 1$$

$$y_1 = 1 - \frac{1}{2}x^2 + \frac{-1}{2}\frac{(-1)}{(4)(3)}x^4$$

$$y_2 = x + \frac{-1}{(3)(2)}x^3 + \frac{-1}{6}\left(\frac{-1}{(5)(4)}\right)x^5$$

So that the general solution is... $\left[y = c_1 y_1 + c_2 y_2\right]$

Where

$$y = c_1\left(1 - \frac{x^2}{2!} + \frac{x^4}{4!} - \ldots\right) + c_2\left(x - \frac{x^3}{3!} + \frac{x^5}{5!} - \ldots\right)$$

The solution to this differential equation should be unique, where the result should map any differential equation. Therefore, we investigate further by considering the Taylor's Series for cos(x) and sin(x) at $x_0 = 0$ (MacLaurin's Series); ideally, this should match the power series solution.

Therefore, we construct the terms of the Taylor series for cos(x) at $x_0 = 0$:

$$f(x) = \cos x \quad f(0) = 1$$

$$f'(x) = -\sin x \quad f'(0) = 0$$

$$f''(x) = -\cos x \quad f''(0) = -1$$

$$f'''(x) = \sin x \quad f'''(0) = 0$$

$$f^4(x) = \cos x \quad f^4(0) = 1$$

Also consider the MacLaurin's series for sin(x) at $x_0 = 0$:

$$g(x) = \sin x \quad g(0) = 0$$

$$g'(x) = \cos x \quad g'(0) = 1$$

$$g''(x) = -\sin x \quad g''(0) = 0$$

$$g'''(x) = -\cos x \quad g'''(0) = -1$$

$$g^4(x) = \sin x \quad g^4(0) = 0$$

$$g^5(x) = \cos x \quad g^5(0) = 1$$

Then, assembling terms, we have

$$F(x) = 1 + 0 \cdot x - \frac{x^2}{2!} + 0 \cdot \frac{x^3}{3!} + \frac{x^4}{4!} + \ldots$$

$$G(x) = 0 + x + 0 \cdot \frac{x^2}{2!} + (-1)\frac{x^3}{3!} + 0 + \ldots$$

So that

$$\cos(x) = F(x) = 1 - \frac{x^2}{2!} + \frac{x^4}{4!} - \dots$$

$$\sin(x) = G(x) = x - \frac{x^3}{3!} + \frac{x^5}{5!} - \dots$$

Comparing this result with general solution of previous page for series solution of

$$y'' + y = 0$$

Note that the solutions from the power series and the solutions from the Taylor's Series yield the analytic power series! This points out that if solved correctly, we should be able to achieve the solution to any differential equation using the power series method that is functionally equivalent to a solution obtained by other means. This is an important result.

The Characteristic Equation

Recall from a previous example, we considered the ODE

$$y'' - y = 0$$

Assuming a solution of the form $y = e^{mx}$ we obtained

$$m^2 - 1 = 0 \qquad y_1 = e^x \quad y_2 = e^{-x}$$

So that the general solution was:

$$y(x) = c_1 e^x + c_2 e^{-x}$$

By inspection, for homogeneous *constant coefficient* ODEs we can obtain the *characteristic equation* by directly substituting in polynomial terms of equivalent degree.

Example

Find the characteristic equation of $y'' + 3y' - 4y = 0$

By inspection, this is $m^2 + 3m - 4 = 0$
$$(m+4)(m-1) = 0$$

Which leads to $m = -4$, $m = 1$, and both roots are real and distinct, so that the solution is

$$\left[y = c_1 e^{-4x} + c_2 e^x \right]$$

Example

Solve $y'' - 7y' = 0$ where we seek a solution $y(x)$

The Characteristic equation is

$$(\lambda - 0)(\lambda - 7) = 0$$

$$\lambda_1 = 0 \quad \lambda_2 = 7$$

$$y = c_1 e^{0x} + c_2 e^{7x} = \left[c_1 + c_2 e^{7x} \right]$$

Example

Solve $y'' + 4y' + 5y = 0$

We then write the characteristic equation

$$\lambda^2 + 4\lambda + 5 = 0$$

$$\lambda = \frac{-4 \pm \sqrt{4^2 - 4(5)}}{2} = \frac{-4 \pm \sqrt{16 - 20}}{2}$$

$$\lambda = \frac{-4 \pm 2i}{2} = -2 \pm i$$

$$\left[y = c_1 e^{-2x} \cos(x) + c_2 e^{-2x} \sin(x) \right]$$

Three Cases for Characteristic Roots

In solving ordinary differential equations with constant coefficients, there are three cases to be considered for the roots of the characteristic polynomial. These are considered here.

Case (i) Real Distinct Characteristic Roots, e.g. m_1, m_2

Then the solution is of the form

$$y = c_1 e^{m_1 x} + c_2 e^{m_2 x}$$

Case (ii) Roots form a Complex Conjugate Pair, e.g.

$$m = a \pm ib \rightarrow \begin{cases} a + ib \\ a - ib \end{cases}$$

Recall $y = d_1 e^{(a+ib)x} + d_2 e^{(a-ib)x}$

From Euler's formula:
$$e^{ibx} = \cos(bx) + i\sin(bx)$$
$$e^{-ibx} = \cos(bx) - i\sin(bx)$$

Then $y = d_1 e^{ax} e^{ibx} + d_2 e^{ax} e^{-ibx}$

$$= d_1 e^{ax}(\cos(bx) + i\sin(bx)) + d_2 e^{ax}(\cos(bx) - i\sin(bx))$$

$$= e^{ax}[(d_1 + d_2)\cos(bx) + i(d_1 - d_2)\sin(bx)]$$

Then, if we assume the constants are in fact a complex conjugate pair, so that $d_1 = \tilde{c}_1 + i\tilde{c}_2$ and $d_2 = \tilde{c}_1 - i\tilde{c}_2$, we can rewrite the solution as

$$y = e^{ax}[(2\tilde{c}_1)\cos(bx) + i(2i\,\tilde{c}_2)\sin(bx)]$$

Because $\tilde{c}_1$ and $\tilde{c}_2$ are arbitrary, we can recast these constants as:

$$(2\tilde{c}_1) \to c_1 \quad \text{and} \quad (2i^2\tilde{c}_2) \to (-2\tilde{c}_2) \to c_2$$

Therefore, the solution becomes

$$\left[y = c_1 e^{ax}\cos(bx) + c_2 e^{ax}\sin(bx) \right]$$

and the solution is real if and only if c_1, c_2 are real valued constants.

Note this was only possible if d_1 *and* d_2 are treated as complex conjugates. Therefore, since we are only interested in real values, we restrict d_1 *and* d_2 and consider them to be a "conjugate pair".

Case (iii) $m_1 = m_2$

This is the case of repeated roots, and can be directly derived using *Reduction of Order* when one solution is known and a second solution must be obtained.

Therefore, for the case of ODEs with constant coefficients, we multiply the second solution by a power in x, so that the two solutions are:

$$y = c_1 e^{m_1 x} + c_2 x e^{m_1 x}$$

Note this directly arises from reduction of order, and this method is valid for linear ODEs with constant coefficients.

Reduction of Order

Reduction of Order is a procedure to find a second solution of the differential equation

$$y'' + P(x)y' + Q(x)y = f(x)$$

for the case where we know at least one solution of the differential equation. Then we assume that

$$y_2 = u(x)y_1(x)$$

Where $u(x)$ "maps" the first solution to a second, linearly independent solution. Applying derivatives of the assumed solution, and placing this back into the differential equation, we can solve for $u(x)$, thereby obtaining the second solution by applying the above equation. The reduction of order process, in compact form, amounts to the following general formulation:

$$\left[y_2(x) = y_1(x) \int \frac{e^{-\int P(x)dx}}{(y_1(x))^2} dx \right]$$

Example

$$y'' - y = 0$$

Step 1: Obtain at a second solution of the equation if we know that $y_1 = e^x$ indeed satisfies the differential equation.

Step 2: Assume $y_2 = u(x)y_1(x) = u(x)e^x$

Then $y_2' = u'(x)e^x + u(x)e^x$

$$y_2'' = u''(x)e^x + u'(x)e^x + u'(x)e^x + u(x)e^x$$
$$= u''(x)e^x + 2u'(x)e^x + u(x)e^x$$

Step 3: Put y_2 and derivative forms back into the differential equation

$$u''(x)e^x + 2u'(x)e^x + u(x)e^x - u(x)e^x = 0$$

$$\Rightarrow e^x\left(u''(x) + 2u'(x)\right) = 0$$

$$e^x \neq 0, \, so: \quad u''(x) + 2u'(x) = 0$$

Step 4: Let $w = u'(x)$; then $w' = u''(x)$

$$\text{So,} \quad w' + 2w = 0$$

Note how the order is reduced, so that we can now solve this reduced order equation by integrating factor $\Rightarrow e^{\int 2dx} = e^{2x}$

$$\frac{d}{dx}\left(e^{2x}w\right) = 0 \cdot e^{2x} = 0 \Rightarrow \int d\left(e^{2x}w\right) = c_1$$

$$e^{2x}w = c_1 \quad w = c_1 e^{-2x} = \frac{du}{dx} \quad u = -\frac{1}{2}c_1 e^{-2x} + c_2$$

So, $u = -\dfrac{1}{2}c_1 e^{-2x} + c_2$

Note that c_1, c_2 are arbitrary constants.

Step 5: Now that we have $u(x)$, recast $y_2 = u(x)y_1$

Therefore $\quad y_2 = \left(-\dfrac{1}{2} c_1 e^{-2x} + c_2 \right) e^x$

And $\quad y_2 = -\dfrac{1}{2} c_1 e^{-x} + c_2 e^x$

Since c_1, c_2 are arbitrary, then choose

$$c_1 = -2c_3 \quad c_2 = 0 \quad y_2 = c_3 e^{-x}$$

Step 6: So the general solution to the differential equation is in agreement with what we achieved for the differential equation previously:

$$\left[y = c_3 e^{-x} + c_4 e^x \right]$$

Note that this approach is followed in a similar manner for *variation of parameters* for deriving a particular solution to the differential equation from the general solution:

$$y'' + P(x)y' + Q(x)y = f(x)$$

For a 2nd order differential equation, variation of parameters is cast in the following manner:

$$y(x) \Rightarrow \begin{cases} u_1' y_1 + u_2' y_2 = 0 \\ u_1' y_1' + u_2' y_2' = f(x) \end{cases}$$

We will discuss this in the next section, where we discuss non-homogeneous solution methods.

Non-Homogeneous Solution Methods

Method of Undetermined Coefficients

This method assumes a form for the particular solution to the ODE, and through substitution, enables one to discern the form of the solution. Note that this method is restricted in its application to ODEs with *constant coefficients*, for particular solutions of limited form. Those forms that can be assumed for this method include:

(i)   Constants:
$$y_p = A$$

(ii) nth degree polynomial:
$$y_p = A_n x^n + A_{n-1} x^{n-1} + ...$$

(iii) exponentials with polynomials:
$$y_p = e^{ax} \left(A_n x^n + ... + A_{n-1} x^{n-1} + ... \right)$$

(iv)   Sines and cosines multiplied by polynomials or exponentials:
$$y_p = e^{\alpha x} \sin \beta x \left(A_n x^n ... \right) + e^{\alpha x} \cos \beta x \left(A_n x^n ... \right)$$

Example

$$y'' - y' - 2y = 4x^2$$
$$y_h = c_1 e^{-x} + c_2 e^{2x}$$
$$f(x) = 4x^2$$

Then we assume the following:

$$y_p = A_2 x^2 + A_1 x + A_0$$
$$y_p' = 2A_2 x + A_1$$
$$y_p'' = 2A_2$$

Where this is substituted into the differential equation, and we solve for the coefficients A_n:

$$2A_2 - (2A_2 x + A_1) - 2(A_2 x^2 + A_1 x + A_0) = 4x^2$$

Match "like" powers

$$-2A_2 = 4 \quad -2A_2 - A_1 = 0$$

$$2A_2 - A_1 - 2A_0 = 0$$

$$A_2 = -2 \quad A_1 = 2 \quad A_0 = -3$$

$$y_p = -2x^2 + 2x - 3$$

The complete solution is

$$y = y_h + y_p = \left[c_1 e^{-x} + c_2 e^{2x} - 2x^2 + 2x - 3 \right]$$

Variation of Parameters

Unlike the method of undetermined coefficients, which is limited to constant coefficient differential equations having specific forms, *variation of parameters* is a powerful method which can be applied to *any* differential equation, whereas the method of undetermined coefficients can only be used for problems involving differential equations with constant coefficients.

To develop this method, assume we have a linear, ordinary, second order differential equation of the form

$$a_n(x)\frac{d^n y}{dx^n} + a_{n-1}(x)\frac{d^{n-1} y}{dx^{n-1}} + \ldots + a_1(x)\frac{dy}{dx} + a_0(x)y = g(x)$$

where we first recast the equation in the form

$$\frac{d^2 y}{dx^2} + P(x)\frac{dy}{dx} + Q(x)y = f(x)$$

We then assume that two solutions to the homogeneous equation, can be obtained

$$\frac{d^2 y}{dx^2} + P(x)\frac{dy}{dx} + Q(x)y = 0$$

Assumed to be y_1, y_2. Therefore, we only need to find the particular solution y_p.

Using variation of parameters, we assume $y_p = u_1 y_1 + u_2 y_2$.

This means that we will be "mapping" y_1, y_2, using the functions u_1, u_2 to provide the solution to the particular part of the differential equation.

Then we take the derivatives of the assumed particular form:

$$y_p = u_1 y_1 + u_2 y_2$$

$$y_p{}' = u_1{}' y_1 + u_1 y_1{}' + u_2{}' y_2 + u_2 y_2{}'$$

$$y_p{}'' = u_1{}'' y_1 + (u_1{}' y_1{}' + u_1{}' y_1{}') + u_1{}' y_1{}'' + u_2{}'' y_2 + (u_2{}' y_2{}' + u_2{}' y_2{}') + u_2 y_2{}''$$

Substituting this into the differential equation

$$y_p{}'' + P(x)y_p{}' + Q(x)y_p = f(x) =$$

$$u_1{}'' y_1 + (u_1{}' y_1{}' + u_1{}' y_1{}') + u_1{}' y_1{}'' + u_2{}'' y_2 + (u_2{}' y_2{}' + u_2{}' y_2{}') + u_2 y_2{}'' +$$

$$P(x)(u_1{}' y_1 + u_1 y_1{}' + u_2{}' y_2 + u_2 y_2{}') + Q(x)(u_1 y_1 + u_2 y_2) = f(x)$$

Factoring terms, we get

$$u_1(y_1''+P(x)y_1'+Q(x)y_1)+u_2(y_2''+P(x)y_2'+Q(x)y_2)+$$
$$u_1''y_1+(2u_1'y_1')+u_2''y_2+(2u_2'y_2')+P(x)(u_1'y_1+u_2'y_2)=f(x)$$

Recognizing the first two terms are zero, then

$$u_1''y_1+(2u_1'y_1')+u_2''y_2+(2u_2'y_2')+P(x)(u_1'y_1+u_2'y_2)=f(x)$$

Which can be recast as

$$\frac{d}{dx}(u_1'y_1+u_2'y_2)+P(x)(u_1'y_1+u_2'y_2)+u_1'y_1'+u_2'y_2'=f(x)$$

Therefore we need to have the following conditions hold for the above equation in order to make this method solvable:

$$y(x) \Rightarrow \begin{cases} u_1'y_1+u_2'y_2=0 \\ u_1'y_1'+u_2'y_2'=f(x) \end{cases}$$

With this condition in place, we can now solve these two equations for the two unknowns

$$u_1',u_2'$$

and then integrate these directly to substitute into the originally proposed form for the particular solution:

$$y_p=u_1y_1+u_2y_2$$

Recalling the definition of the Wronskian, it can be seen that the conditions for setting up variation of parameters using the following Wronskians is

$$W = \begin{vmatrix} y_1 & y_2 \\ y_1' & y_2' \end{vmatrix} \qquad W_1 = \begin{vmatrix} 0 & y_2 \\ f(x) & y_2' \end{vmatrix} \qquad W_2 = \begin{vmatrix} y_1 & 0 \\ y_1' & f(x) \end{vmatrix}$$

$$u_1' = \frac{W_1}{W} = \frac{-y_2 f(x)}{W} \qquad \text{and} \qquad u_2' = \frac{W_2}{W} = \frac{y_1 f(x)}{W}$$

Where integration of these is performed as stated above. As we will discuss later, this comes from application of *Cramer's Rule*.

Substitution Methods

On occasion, there are forms of differential equations that lend themselves to be solved using a substitution method that affords a solution. One such equation is the *Cauchy–Euler differential equation*, which as the form

$$x^2 y'' + axy' + by = 0$$

where a and b are constants, the substitution of $|x| = e^z$ or $z = \ln x$ sufficiently simplifies the equations so that the original equation can be reduced to an equation with constant coefficients that can be directly solved.

Example

$$x^2 y'' - xy' + y = x^5$$

We identify that this is a Cauchy–Euler equation.

Therefore, we call for a substitution where

$$|x| = e^z \quad \text{or} \quad z = \ln x$$

Then the first derivative becomes

$$\left[y' = \frac{dy}{dx} = \frac{dy}{dz}\frac{dz}{dx} = \frac{dy}{dz}\frac{1}{x} \right]$$

Then we can determine the second derivative using the first derivative:

$$y'' = \frac{d}{dx}\left(\frac{dy}{dx}\right) = \frac{d}{dx}\left(\frac{dy}{dz}\frac{1}{x}\right)$$

$$y'' \Rightarrow \frac{d}{dx}\left(\frac{dy}{dz}\right)\cdot\frac{1}{x} + \frac{dy}{dz}\cdot\left(\frac{-1}{x^2}\right)$$

Rewrite $\dfrac{d}{dx} \rightarrow \dfrac{dz}{dx}\dfrac{d}{dz}$ then using the chain rule

$$y'' \Rightarrow \frac{dz}{dx}\frac{d}{dz}\left(\frac{dy}{dz}\right)\cdot\frac{1}{x} + \frac{dy}{dz}\left(-\frac{1}{x^2}\right)$$

$$y'' \Rightarrow \frac{dz}{dx}\frac{d^2y}{dz^2}\cdot\frac{1}{x} + \frac{dy}{dz}\left(-\frac{1}{x^2}\right)$$

$$y'' \Rightarrow \frac{1}{x}\frac{d^2y}{dz^2}\frac{1}{x} + \frac{dy}{dz}\left(-\frac{1}{x^2}\right)$$

$$\left[y'' = \frac{1}{x^2}\frac{d^2y}{dz^2} + -\frac{1}{x^2}\frac{dy}{dz} \right]$$

Note: you must convert the non-homogeneous form also!

The homogeneous form:

$$x^2 y'' - xy' + y = 0$$

Becomes:

$$\frac{d^2y}{dz^2} - \frac{dy}{dz} - \frac{dy}{dz} + y = 0 \qquad or \qquad y'' - 2y' + y = 0$$

Now solve via traditional method.

To obtain the homogeneous solution, the characteristic polynomial is then, by inspection

$$\lambda^2 - 2\lambda + 1 = 0;$$

$$\therefore (\lambda - 1)^2 = 0 \qquad \lambda_1 = 1,\ \lambda_2 = 1$$

$$y_n(z) = c_1 e^z + c_2 z e^z$$

The particular solution is since $z = \ln x \quad e^z = x$ then $x^5 = e^{5z}$

Then for (y(z)):

$$y'' - 2y' + y = e^{5z}$$

In this case, since we transformed the equation into one with constant coefficients, we can make use of the method of undetermined coefficients, where in this application we assume the following form for the particular solution:

$$y_p = Ae^{5z}$$

$$y'_p = 5Ae^{5z} \quad y''_p = 25Ae^{5z}$$

$$\left(25Ae^{5z} - 2\left(5Ae^{5z}\right) + Ae^{5z} = e^{5z}\right)$$

$$25A - 10A + A = 1$$

$$16A = 1$$

$$A = \frac{1}{16} \quad y_p = \frac{1}{16}e^{5z}$$

Then, we recast the solutions using $z = \ln x$ according to

$$y(x) = c_1 x + c_2 \ln x \cdot x + \frac{1}{16}\exp(5\ln x)$$

Or

$$\left[y(x) = c_1 x + c_2 x \ln x + \frac{x^5}{16} \quad x > 0 \right]$$

As an alternative, we could also find a particular solution using variation of parameters...

Therefore, we assume $y_p = u_1 y_1 + u_2 y_2$

And accordingly:

$$y(z) \Rightarrow \begin{cases} u_1' y_1 + u_2' y_2 = 0 \\ u_1' y_1' + u_2' y_2' = e^{5z} \end{cases}$$

$$y_1 = e^z \quad y_2 = ze^z$$
$$y_1' = e^z \quad y_2' = e^z + ze^z$$

$$u_1' e^z + u_2' ze^z = 0$$
$$u_1' e^z + u_2' \left(e^z + ze^z\right) = e^{5z}$$

$$u_1' = -u_2' z$$

$$-u_2' ze^z + u_2' e^z + u_2' ze^z = e^{5z}$$
$$u_2' = e^{4z}$$

$$\frac{du_2}{dz} = e^{4z} \qquad u_2(z) = \int e^{4z} dz = \frac{1}{4} e^{4z} + \tilde{c}$$
$$\frac{du_1}{dz} = -ze^{4z} \quad \int du_1 = \int -ze^{4z} dz$$

$$u = -z \qquad dV = e^{4z} dz$$
$$du = -dz \quad V = \frac{1}{4} e^{4z}$$

$$u_1 = \frac{-z}{4} e^{4z} + \int \frac{1}{4} e^{4z} dz \Rightarrow \frac{-z}{4} e^{4z} + \frac{1}{16} e^{4z}$$

$$u_1 = e^{4z}\left(\frac{-z}{4} + \frac{1}{16}\right); \quad u_2 = \frac{e^{4z}}{4}; \quad y_p = u_1 y_1 + u_2 y_2$$

$$y_p = e^{4z}\left(\frac{-z}{4} + \frac{1}{16}\right) e^z + \frac{e^{4z}}{4} ze^z$$

$$y_p = \frac{-ze^{5z}}{4} + \frac{1}{16} e^{5z} + \frac{ze^{5z}}{4}$$

$$y_p = \frac{1}{16} e^{5z}$$

$$y(z) = y_h + y_p = c_1 e^z + c_2 z e^z + \frac{1}{16} e^{5z}$$

$$x = e^z \quad z = \ln x \quad x > 0$$

$$\left[y(x) = c_1 x + c_2 x \ln x + \frac{x^5}{16} \quad x > 0 \right]$$

Note: this is the same result as we obtained previously, as should be expected.

Substitution methods can be found for several differential equations, and the simplification that results from these methods is important in affording a more direct solution.

NUCLEAR APPICATION: Flow Induced Vibration Analysis

Nuclear fuel elements in reactor cores are often packaged as cylinder bundles supported by end plates, support grids, bearing pads, and other structures. Depending upon their geometry, flexural rigidity, cross flow and axial flow patterns, the impact of Flow Induced Vibration (FIV) and associated "fretting wear" related to component lifetimes must be considered in system design, since FIV can severely impact the design life of nuclear reactors and related components. In particular, nuclear fuel bundles constitute "highly non-linear and ill-defined systems" that change mechanical characteristics as a result of heat transfer and irradiation characteristics.

When damping in nuclear fuel cooled with high speed water for convective heat removal becomes overwhelmed by flow induced oscillations, mechanical failures can occur. An entire series of courses could be devoted to this subject, however, FIV vibration analysis is essentially related to the solution of a complex set of spring mass systems with varying levels of damping.

For illustration, let's consider a single fuel rod in a nuclear reactor with water coolant, where the coolant moves over the fuel at a high velocity to facilitate heat removal. Fuel rod vibration can be characterized as a spring mass system. A Fixed-Hinged system is shown below.

A "Fixed-Hinged" fuel rod of length L_o

The natural frequency in (Hz) of a fuel rod can be computed as follows:

$$f_n = \frac{C_n^2}{2\pi L_o^2} \sqrt{Ey \frac{Ix}{m_{\text{length}}}}$$

Where C_n is the modal vibration number for a set of end conditions; for a fixed-hinged rod, the dominant mode for vibration analysis is the fundamental mode, $n = 1$, $C_1 = 3.927$; for second and third mode numbers, $C_2 = 7.069$, $C_3 = 10.210$. The length L_o is the characteristic length of the rod (e.g. the distance in cm between grid spacer and end), Ey is Young's modulus in $g/(\text{cm s}^2)$, Ix is the area moment of inertia (resistance to bending) in cm^4, m_{length} is the mass of the fuel rod per unit length (g/cm).

For a round cylinder, the moment of inertia is based on rod radius to the fourth power:

$$Ix = \frac{1}{4} \pi (r_o)^4$$

To determine the amplitude of vibration of this rod at the point of maximum deflection, this is governed by Newton's law, the sum of the forces equals mass times acceleration, so that for a spring with damping:

$$\Sigma F = m \frac{d^2 x}{dt^2} \rightarrow m\ddot{x} = -C_{sp}x - B_d\dot{x}$$

The equation for the unforced system then becomes:

$$\ddot{x} + \frac{C_{sp}}{m}x + \frac{B_d}{m}\dot{x} = 0$$

where C_{sp} is the rod effective spring stiffness in (g/s²) per unit length, and B_d is the velocity damping coefficient in g/s. The damped vibration with no outside forcing is:

$$x_n[t] = x_o \mathrm{Exp}(-\zeta\omega_n t)\mathrm{Cos}[(\omega_n t\sqrt{1-\zeta^2}) - \varphi]$$

where the natural frequency is

$$(2\pi f_n) = \omega_n = \sqrt{\frac{C_{sp}}{m}}$$

and the damping factor and phase angle are

$$\zeta = \frac{B_d}{2\sqrt{mC_{sp}}} \quad \text{and} \quad \varphi.$$

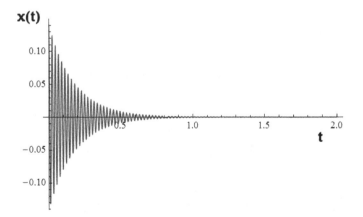

Plot of amplitude vs. time for fundamental mode fuel rod at the center point of maximum deflection, $x_n[t] = 0.140447e^{-5.59074t} \mathrm{Sin}[279.481\,t]$, $\omega_n = 279.54$ rad/s, $\zeta = 0.02$

The system can then be solved for an external forcing function F_{ext} introduced into the system, such as caused by flow, and this becomes:

$$\ddot{x} + \frac{C_{sp}}{m} x + \frac{B_d}{m} \dot{x} = \frac{F_{ext}}{m}$$

Where the forcing function F_{ext} can follow a periodicity of the flow induced vibration.

All the methods of solving ordinary differential equations briefly reviewed here are needed in solving partial differential equations later in the text. Moreover, in general, as will be evident, an alternative is to solve the differential equation using power series methods; a means of addressing potential singularities in "deleted neighborhoods" around "regular singular points" of expansion for the power Frobenius series is important to address.

Chapter 6

Applications of Power Series

Power series are useful tools that can be used to expand other functions, solve equations, provide for assessment of intervals of convergence, used as trial functions, and are applied in all areas of engineering. Included in this discussion are Taylor's Series, which are extremely important in numerical approximations. We present here a brief discussion of Power Series with several applications.

Power Series

A *Power Series* is an infinite series that is summed with increasing powers of x. The general form is:

$$\sum_{m=0}^{\infty} a_m (x - x_0)^m = a_0 + a_1 (x - x_0) + a_2 (x - x_0)^2 ...$$

Where
$a_m = m^{th}$ term coefficient, which can be positive, negative, or zero,
$x_0 = $ power series center value (expansion point), and
$x = $ independent variable

Special Case for $x_0 \to 0$ then

$$\sum_{m=0}^{\infty} a_m x^m = a_0 + a_1 x + a_2 x^2 ...$$

The Truncated series is $S_n(x) = \sum_{m=0}^{n} a_m (x - x_0)^m$

The remainder of the series is $R_n(x) = \sum_{m=n}^{\infty} a_m (x - x_0)^m$

The complete power series is

$$S(x) = \sum_{m=0}^{\infty} a_m (x - x_0)^m = S_n(x) + R_n(x)$$

Limits, Continuity, and Derivatives of Functions

A real valued function is analytic at a point if it can be represented by a power series.

Also, for real valued functions, limits must exist from both sides for continuity/differentiability of a function. Analytic functions must preserve mixed partial derivatives as follows:

$$\frac{\partial^2 f}{\partial x \partial y} = \frac{\partial^2 f}{\partial y \partial x} \qquad \frac{\partial}{\partial y} \frac{\partial f}{\partial x} = \frac{\partial}{\partial x} \frac{\partial f}{\partial y}$$

If this criteria is not met, then the function is *not* analytic. This is effectively analogous to the Cauchy–Riemann Equations as was applied to Complex Functions.

Interval of Convergence

All values of x identified for which the series converges, where

$$\left(|x - x_0| < R \right)$$

This is most often tested/determined using the *Ratio Test*.

Ratio Test

The Ratio Test is applied on the coefficient terms

$$L = \lim_{n \to \infty} \left| \frac{a_{m+1}}{a_m} \right| < 1 \quad \text{for convergence}$$

If $L \to \infty$ then only convergent at x_0.

If $L > 1$, $S(x)$ diverges; If $L < 1$, $S(x)$ converges to a stable limit.

If $L = 1$, no information is available and we must use another test

Often, when the Ratio Test yields no information, viable substitutes include the root test, integral test, comparison test, or alternating series test.

Root Test

$$\text{Find} \quad \lim_{n \to \infty} \sqrt[m]{|a_m|}$$

The series converges if the limit exists and is non-zero.

Integral Test

Substituting $a_n \to a_x$ and is to be used if/where $a_x > 0$ and decreases for $(x > \xi)$

$$\int_{\xi}^{\infty} a_x dx = exists(finite)$$

Comparison Test

A comparison is made to the

$$\text{“P-series”} \rightarrow 1 + \frac{1}{2^p} + \frac{1}{3^p} + \frac{1}{4^p}$$

which converges for $p > 1$, and diverges otherwise

Alternating Series Test

Convert the series coefficient to a continuous function $f(x)$, and if that function decreases for $x > 0$ and the x-axis is an asymptote, then this Alternating Series is *convergent*.

This is also often presented as considering $0 < a_{k+1} < a_k$ for all "k" integers and a_n has an asymptote on the x-axis.

Example

Find the interval (radius) of convergence of the power series…

$$\sum_{m=0}^{\infty} (-1)^{m+1} \frac{1}{m!} x^m$$

Applying the ratio test

$$\lim_{n \to \infty} \left| \frac{\dfrac{1}{(n+1)!}}{\dfrac{1}{n!}} \right| = \lim_{n \to \infty} \left| \frac{1}{n+1} \right| \Rightarrow 0 < 1$$

$\therefore$ the series converges by the ratio test.

Since the limit is identically <u>zero</u> for every value of x, then the series is "absolutely convergent" for all real numbers.

For the Radius of Convergence, considering the standard power in each term is x^m, then

$$R \equiv \text{radius of convergence is } |x| < \frac{1}{\lim_{n \to \infty} \left| \dfrac{a_{n+1}}{a_n} \right|}$$

And in this example,

$$|x| < \frac{1}{0} \to \infty$$

so that we conclude that $-\infty < x < \infty$. Therefore, this is a convergent series for all real values of x.

Example

Find the interval (radius) of convergence of the power series…

$$\sum_{m=0}^{\infty} (-1)^{m+1} \frac{1}{m} x^m$$

Applying the ratio test:

$$\lim_{n \to \infty} \left| \frac{\dfrac{1}{n+1}}{\dfrac{1}{n}} \right| \Rightarrow \lim_{n \to \infty} \left| \frac{n}{n+1} \right|$$

assuming "n" is large and non-zero

$$\Rightarrow \lim_{n\to\infty} \left| \frac{n/n}{n/n + 1/n} \right| \Rightarrow \lim_{n\to\infty} \left| \frac{1}{1 + 1/n} \right| \Rightarrow 1 = 1$$

Note this reveals nothing by the ratio test.

Since the series is alternating, we can apply the alternating series test, and the function

$$f(x) = \frac{1}{x} \quad \text{decreases for } x > 0,$$

with the x-axis as an asymptote, this alternating series converges.

(Note also, since $0 < a_{k+1} < a_k$ for all "k" integers, and the x-axis serves as an asymptote then this converges by the alternating series test. Note that as given, this series is the *Alternating Harmonic Series* which converges by the *Alternating Series Test*.)

Example

Is the series $\displaystyle\sum_{m=0}^{\infty} \frac{1}{m}$ convergent?

Based on the P-series test, it is a divergent *P*-series.

Example

Examine the series using the Integral Test:

$$\sum_{m=1}^{\infty} \frac{1}{\sqrt{2m+1}} x^m = \frac{x^1}{\sqrt{3}} + \frac{x^2}{\sqrt{5}} + \frac{x^3}{\sqrt{7}} + \dots$$

Applying the integral test:

$$\lim_{b \to \infty} \int_1^b \frac{1}{\sqrt{2x+1}} dx = \lim_{b \to \infty} (\sqrt{2x+1})\, |_1^b \to \infty$$

Therefore, the series is divergent.

Example

Find radius of convergence R for

$$\sum_{m=0}^{\infty} \frac{(-1)^m}{8^m} x^{3m} = 1 - \frac{x^3}{8} + \frac{x^6}{64} - \frac{x^9}{512} + \dots$$

Applying the ratio test:

$$\lim_{n \to \infty} \left| \frac{\frac{1}{8^{n+1}}}{\frac{1}{8^n}} \right| = \lim_{n \to \infty} \left| \frac{8^n}{8^n 8} \right| \Rightarrow \frac{1}{8}$$

Now, considering the standard power in each sequential term of the series is x^{3m} (compared to the standard first power x^m), then we must use a *third power* in the inequality to determine the Radius of Convegence:

$$|x|^3 < R = \frac{1}{\frac{1}{8}} = 8$$

In this case, simplifying using a cube root, the series is convergent for

$$|x| < 2$$

Always include the absolute value sign because it was required in the ratio. Therefore:

$$-2 < x < 2$$

Operations on Power Series

Standard numerical operations can be performed on all Power Series, including

- Addition
- Subtraction
- Multiplication
- Division (provided continuity is preserved).

Within the radius of convergence R, derivatives and integrals are also permissible, but enough terms must be considered for appropriate accuracy, continuity, and convergence within a specific tolerance.

We will explore more on power series operations when we apply them in solving several differential equations later in the text.

Taylor's Series

A *Taylor's Series* is an expansion of a continuously differentiable function into a *finite series* with a *remainder term*. This was first published by the English mathematician Brook Taylor in 1715, and can be derived in the following manner:

Given $f^n(x)$ is the "nth" derivative of the continuously differentiable function $f(x)$ ...

$$\int_a^x f^n(x_0)dx_0 = \int_a^x \frac{d^n f}{dx_0}dx_0 = \int_a^x d^n f$$

This yields $\Rightarrow \left(f^{n-1}(x) - f^{n-1}(a)\right)$

If we integrate the above equation again in the same way...

$$\int_a^x \int_a^x f^n(x_0)dx_0 dx_0 = \int_a^x \left[f^{n-1}(x_0) - f^{n-1}(a)\right]dx_0$$

Yields $\Rightarrow \left(f^{n-2}(x) - f^{n-2}(a)\right) - \left(f^{n-1}(a)(x-a)\right)$

Integrating yet again:

$$\int_a^x \int_a^x \int_a^x f^n(x_0)\,dx_0\,dx_0\,dx_0$$

$$\Rightarrow \left(f^{n-3}(x) - f^{n-3}(a)\right) - \left(f^{n-2}(a)(x-a)\right) - \left(f^{n-1}(a)\frac{(x-a)^2}{2}\right)$$

So that by integrating for the *n*th time we obtain…

$$\int_a^x \dots \int f^n(x_0)(dx_0)^n =$$

$$f(x) - f(a) - (x-a)f'(a) - \frac{(x-a)^2}{2!}f''(a) - \dots - \frac{(x-a)^{n-1}}{(n-1)!}f^{n-1}(a)$$

We note that the above equation is exact, since no terms or simplifications have occurred! Also, if one can find the Taylor's series for a function, it is stable and convergent.

Now, we solve the last equation for $f(x)$ to yield the single variable Taylor's series formula:

$$f(x) = f(a) + (x-a)f'(a) + \frac{(x-a)^2}{2!}f''(a) + \dots + \frac{(x-a)^{n-1}}{(n-1)!}f^{n-1}(a) + R_n$$

Where the truncated series remainder term is

$$R_n = \int_a^x \dots \int f^n(x_0)(dx_0)^n$$

From the Mean Value Theorem of Calculus, we can write

$$\int_a^x g(x)\,dx = (x-a)g(\xi), \text{ where } \xi \equiv \text{a mean value on } [a,x]$$

$$\left\{ R_n = \frac{(x-a)^n}{n!} f^n(\xi) \right\}$$

In summary, Taylor's series expansion of a function ("well behaved" in neighborhood of "a")

$$f(x) = f(a) + f'(a)(x-a) + \frac{f''(a)}{2!}(x-a)^2 + ...$$

Or in series notation, $f(x) = \sum_{n=0}^{\infty} \frac{f^n(a)}{n!}(x-a)^n$

Note: If we "run out" of derivatives, as in a Taylor series of a polynomial, we exactly reproduce $f(x)$.

Example

Consider $f(x) = x^2 - 2x$; find the Taylor's series at $a = 1$.

Taking derivatives

$$f'(x) = 2x - 2 \quad f''(x) = 2 \quad f'''(x) = 0$$

By Taylor's Theorem, expand at $a = 1$

$$f(x) \approx f(a) + f'(a)(x-a) + \frac{f''(a)}{2!}(x-a)^2$$

$$f(x) = (1^2 - 2) + (2(1) - 2)(x-1) + \frac{2}{2!}(x-1)^2$$

$$f(x) = -1 + 0 + (x-1)^2 = -1 + 0 + x^2 - 2x + 1$$

$$f(x) = x^2 - 2x$$

Note that in this case, the complete Taylor series of a polynomial is the function itself! This is the immutable truth of the uniqueness of the Taylor's series for all functions. This will be apparent as we move forward in solving all types of problems; as we will find out, the Taylor's series can be used to form the basis for all differential equations, differencing schemes, and thereby most every application in engineering and physics!

A special case of the Taylor's series is the MacLaurin's Series, where the Taylor's series applied at an expansion point of zero.

Example

Find $g(x)$, the third order MacLaurin's Series of $f(x) = \sin(x/2)$ and compare this with the Taylor formula expanded from $a = 2\pi$.

Applying the Taylor formula, we get $g_1(x) = (x/2) - (x^3/48)$.

Plotting this alongside the original function gives us insight into how well the series approximates the function:

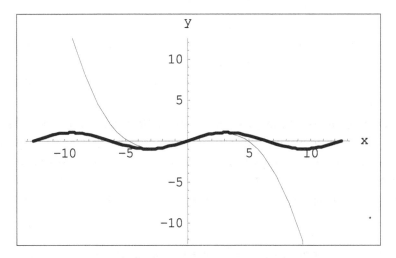

Figure: Plot of $f(x)$ and $g_1(x)$

From the figure, it is clear that the series $g_1(x)$ is quite reasonable in the local neighborhood of the point of expansion of the series, but indeed deviates quickly beyond the interval $[-3,3]$.

For comparison, the series expanded from $a = 2\pi$ yields

$$g_2(x) = ((x - 2\pi)/2) + ((x - 2\pi)^3/48)$$

The plot of this function is:

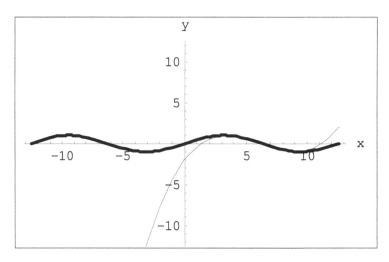

Plot of $f(x)$ and $g_2(x)$

As evident in the figure, the effect of shifting the series gave us a more accurate approximation for values of x between 3 and 9. By not having additional terms in the series, which are effectively truncated and become part of the *truncation error*, the accuracy of the Taylor's series suffers. Truncation error is that part of the series that is ignored, or truncated.

To capture better accuracy over a wider interval, we must consider more terms for the series. Since the fourth derivative of $f(x) = \sin(x/2)$ is zero at $a = 2\pi$, we go on to the 5th derivative:

$$g_3(x) = ((x - 2\pi)/2) + ((x - 2\pi)^3 / 48) - ((x - 2\pi)^5 / 3840)$$

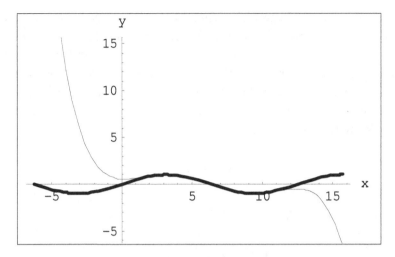

Figure: Plot of $f(x)$ and $g_3(x)$

The effect of the additional term is readily visible in the plot of $g_3(x)$, and the interval where we observe good accuracy of the series is much broader.

It would be convenient to know how accurately our Taylor's series function is; it turns out that this is readily available to us.

Approximating the Truncation Error of a Taylor's Series

In general the truncation error resulting from truncating the series at the nth order is approximately the size of the term out at the $(n+1)^{st}$ order. This is because as higher and higher order series terms are considered, these higher order terms typically add or subtract vanishingly small corrections to the series with increasing order.

Therefore, one can gain an appreciation for the magnitude of the truncation of the Taylor's series by readily evaluating the magnitude of the $(n+1)^{st}$ term, since all higher order terms, while forming a sum, tend to be smaller and smaller with increasing n. Of course, it is important to note that for this logic to hold, we take advantage of the fact that the Taylor's series is convergent for all real values of x. At this point, we return to the example of the MacLaurin's series of

$$f(x) = \sin(x/2)$$

The (n+1)st term of the truncated series

$$g_1(x) = (x/2) - (x^3/48) \quad \text{is} \quad \varepsilon_T = ((x)^5/3840).$$

Using this as our estimator, we can estimate the truncation error; it is compared to the true error in the table below:

Comparison of the Actual Truncation Error with Estimated Error using the $(n+1)^{st}$ term from the Taylor Series.

x	Actual Error using function	Estimated Error Using (n+1)st term
-5.0	-0.702638811	0.813802083
-4.0	-0.24263076	0.266666667
-3.0	-0.059994987	0.06328125
-2.0	-0.008137651	0.008333333
-1.0	-0.000258872	0.000260417
0.0	0	0
1.0	0.000258872	0.000260417
2.0	0.008137651	0.008333333
3.0	0.059994987	0.06328125
4.0	0.24263076	0.266666667
5.0	0.702638811	0.813802083

Multivariate Taylor's Series

Considering $z = f(x, y)$, in a similar manner as for a single variable function (without explicitly deriving it), we can expand Taylor's series from a point $x = a$, $y = b$ on surface $z = f(a,b)$.

$$f(x, y) = f(a,b) + (x - a)\frac{\partial f}{\partial x}\bigg|_{(a,b)} + (y - b)\frac{\partial f}{\partial y}\bigg|_{(a,b)}$$

$$+ \frac{1}{2!}\left[(x-a)^2\frac{\partial^2 f}{\partial x^2}\bigg|_{(a,b)} + 2(x-a)(y-b)\frac{\partial^2 f}{\partial x \partial y}\bigg|_{(a,b)} + (y-b)^2\frac{\partial^2 f}{\partial y^2}\bigg|_{(a,b)}\right]$$

$$\frac{1}{3!}\left[(x-a)^3\frac{\partial^3 f}{\partial x^3}\bigg|_{(a,b)} + 3(x-a)^2(y-b)\frac{\partial^3 f}{\partial x^2 \partial y}\bigg|_{(a,b)}\right.$$

$$\left. + 3(x-a)(y-b)^2\frac{\partial^3 f}{\partial x \partial y^2}\bigg|_{(a,b)} + (y-b)^3\frac{\partial^3 f}{\partial y^3}\bigg|_{(a,b)} \cdots\right]$$

Note: this formulation can be generally applied to as many variables as needed. Taylor's series are very powerful tools that can be applied in a number of ways, and their use will recur throughout this text.

Mathematica Example: Taylor's Series

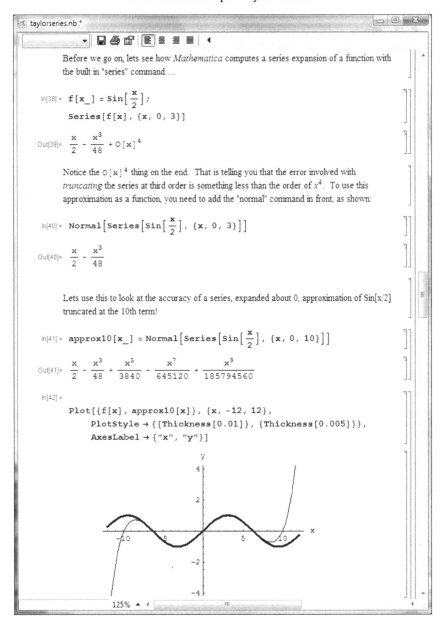

Pade Approximations

Sometimes it is necessary to have a quotient of series expansions. Such an idea was first proposed by French mathematician Henri Eugène Padé (1863–1953), and requires that a function $f(x)$ and its derivatives be continuous at the expansion point, typically $x = 0$.

$$f(x) \approx \frac{P(x)}{Q(x)}$$

where $P(x)$ and $Q(x)$ are Taylor's Series potentially of different order. A rapidly convergent Pade series approximation for when $x < 10^{-4}$ is as follows:

$$g(x) = \left(\frac{1-x}{x} \right) \quad \rightarrow \quad g(x) \approx 1 + \sum_{k=1}^{\infty} (-1)^k \left(\frac{x^k}{(k+1)!} \right)$$

The above expression can occur often in nuclear engineering when dealing with small exponential arguments, as in Bateman equations for isotope production. Therefore, particularly when programming these types of expressions in numerical algorithms, these types of expressions are important.

Chapter 7

Methods of Solving Differential Equations with Variable Coefficients

We already discussed constant coefficients and methods of solutions. Now we discuss more *general solution methods and applications*. First we'll review analytic functions and points of interest.

Analytic Functions

Polynomials, sines, cosines, exponential functions are analytic for all real numbers. Recall a function is *analytic* at x_0 if it has a Taylor's series that exists in the *neighborhood* of some real value x_0, where the Taylor's series, as discussed previously, is given by:

$$\left[\sum_{n=0}^{\infty} \frac{f^n(x)(x-x_0)^n}{n!} \right]$$

Now consider a differential equation with variable coefficients of homogeneous form, and in particular the second order equation

$$b_2(x)y'' + b_1(x)y' + b_0(x)y = 0$$

First, we begin to rearrange this equation by dividing through by $b_2(x)$ and in doing so, we assume $b_2(x) \neq 0$. Therefore:

$$y'' + \frac{b_1(x)}{b_2(x)} y' + \frac{b_0(x)}{b_2(x)} y = 0$$

And we recast according to $\left(p(x) = \dfrac{b_1(x)}{b_2(x)}; \quad q(x) = \dfrac{b_0(x)}{b_2(x)} \right)$

Where the general homogeneous ordinary differential equation is

$$y'' + p(x)y' + q(x)y = 0$$

Also, if a *non-homogeneous equation*, we state this as

$$y'' + p(x)y' + q(x)y = f(x)$$

Classification of Points

We wish to evaluate the solution to the differential equation

$$y'' + p(x)y' + q(x)y = 0$$

assuming we have variable coefficients in the neighborhood of a point x_0. In order to obtain the solution, the point must be classified as either an ordinary point, a regular singular point, or an irregular singular point.

Ordinary Points

x_0 is an "ordinary point" of the differential equation if both $p(x)$ and $q(x)$ are *analytic* at x_0 and can be evaluated at $p(x)$ and $q(x)$ without singularities, and these functions are continuous and differentiable in the neighborhood surrounding x_0.

Regular Singular Points

x_0 is classified as a "regular singular point" of the differential equation above if x_0 can be evaluated at $p(x)$ or $q(x)$ to yield a singular value, and $(x - x_0)p(x)$ and $(x - x_0)^2 q(x)$ are *both analytic* at the point x_0.

If these criteria hold true, then a solution can be found in a "deleted neighborhood" surrounding the regular singular point x_0 for some R radius of convergence

$$(x_0 - R, x_0 + R)$$

with the exception of $x = x_0$.

Therefore, a "deleted neighborhood" $0 < |x - x_0| < R$ of x_0 consists of all numbers $(x_0 - R, x_0 + R)$ except at the specific point where $x = x_0$.

Irregular Singular Points

x_0 is classified as an "irregular singular point" of the differential equation if it cannot be found to qualify as a "regular" singular point.

Example

Determine if $x_0 = 0$ is an ordinary point of $y''-3xy'+4y = 0$

Answer: $p(x) = -3x$

                      $q(x) = 4$

where we note both of these are technically polynomials; these functions are analytic everywhere, including $x_0 = 0$, so that $x_0 = 0$ is an *ordinary point* of the differential equation.

Example

Are $x_o = 1$, $x_o = 3$ ordinary points of $(x^2 - 9)y'' + y = 0$?

Divide the equation through by $(x^2 - 9)$; the differential equation then becomes

$$y'' + \frac{0}{\left(x^2 - 9\right)}y' + \frac{1}{\left(x^2 - 9\right)}y = 0$$

$$p(x) = 0 \quad q(x) = \frac{1}{(x^2 - 9)} = \frac{1}{(x+3)(x-3)}$$

At $x_0 = 1$ $p(x)$ and $q(x)$ are *analytic*, so that $x_0 = 1$, is an *ordinary point* of the differential equation. However, at $x_0 = 3$ $p(x) = 0$, and $q(x)$ is *singular*, and therefore *not analytic* there.

So, $x_0 = 3$ is a *singular point*, and we must then determine if it is a "regular" or "irregular" singular point by following the procedure:

$$[(x-3)p(x)] = 0$$

$$[(x-3)^2 q(x)] = \frac{(x-3)(x-3)}{(x^2-9)} = \frac{(x-3)(x-3)}{(x+3)(x-3)} = \frac{(x-3)}{(x+3)}$$

Based on the this, $x_0 = 3$ is determined to be a *"regular" singular point* of the differential equation.

Power Series Solutions About an Ordinary Point

Again consider the standard general form for the differential equation

$$y'' + p(x)y' + q(x)y = 0$$

If x_0 is an *ordinary point* of the differential equation, then the general solution in an interval containing the point x_0 is given by a power series:

$$\left[y = \sum_{n=0}^{\infty} a_n (x - x_0)^n \equiv a_0 y_1(x) + a_1 y_2(x) \right]$$

where a_0, a_1 are arbitrary constants, and y_1, y_2 are linearly independent analytic functions at x_0.

Power Series Solution Method in a Stepwise Approach:

The power series solution to the differential equation can be derived using a step-wise approach:

Step (i). Substitute the power series form of y, y', y'' into the differential equation and series expansions of $p(x)$ and $q(x)$ (as applicable) in the equation.

Step (ii). Collect like powers of $(x - x_0)$ and set each collected coefficient = 0.

Step (iii). Solve the resulting equations for a_n (n = 2, 3, 4, ...) to effectively yield the power series solutions multiplied by independent series terms a_0, a_1 to produce two linearly independent solutions $(a_0 y_1 + a_1 y_2)$.

Helpful Axioms of Applying Power Series Method

Axiom 1

If $x_0 \neq 0$, the algebra can be simplified by using a change of variables in the differential equation. The parameter t can be introduced, where

$$t = (x - x_o)$$

allowing for solution of a power series in terms of t about the ordinary point $t = 0$. This can then be converted back in terms of x by back-substitution. An applicable radius of convergence may be obtained by applying the ratio test or other suitable convergence test.

Axiom 2

If $p(x)$ and $q(x)$ are quotients of polynomials, it may be practical to multiply through the differential equation by the lowest common denominator, collecting terms incorporating polynomial coefficients.

Axiom 3

For non-homogeneous solutions, solutions obtained from the two homogeneous equation series solutions y_1, y_2 can be used with application of the method of variation of parameters to obtain a particular solution

These "helpful axioms" are also applicable when solving ODE's at regular singular points using the Method of Frobenius...

However, we need two solutions to form a basis for a general solution to the second order ODE...Just like you need an 'x-axis' and a 'y-axis' to define a point on the x-y plane, you need *two* basis functions to span the solution set of a second order ODE.

Example

Solve Legendre's in the neighborhood of $x_0 = 0$.

Consider Legendre's differential equation

$$\left(1-x^2\right)y''-2xy'+m(m+1)y = 0$$

where $y(1) = 1$ and m is a positive integer.

Then from the discussion, we have

$$p(x) = \frac{-2x}{\left(1-x^2\right)} \quad q(x) = \frac{m(m+1)}{\left(1-x^2\right)}$$

And we determine that $x_0 = 0$ is an ordinary point. Since there are non-constant coefficients in this differential equation, we apply the power series method. Noting *Axiom 2* from the "helpful Axioms" we begin with:

$$\left(1-x^2\right)y'' - 2xy' + m(m+1)y = 0$$

And by the power series method:

$$x_0 = 0 \quad \text{so,} \quad y = \sum_{n=0}^{\infty} a_n (x-0)^n = \sum_{n=0}^{\infty} a_n x^n$$

We then take derivatives:

$$y = a_0 + a_1 x + a_2 x^2 + a_3 x^3 + a_4 x^4 + \ldots$$

$$y' = a_1 + 2a_2 x + 3a_3 x^2 + 4a_4 x^3 + \ldots$$

$$y'' = 2a_2 + 6a_3 x + 12a_4 x^2 + \ldots$$

Substituting into the equation, we obtain:

$$\left(1 - x^2\right)\left(2a_2 + 6a_3 x + 12a_4 x^2 + \ldots\right) - 2x\left(a_1 + 2a_2 x + 3a_3 x^2 + \ldots\right) +$$
$$m(m+1)\left(a_0 + a_1 x + a_2 x^2 + a_3 x^3 + \ldots\right) = 0$$

And performing some algebra:

$$\left(\left(2a_2 + 6a_3 x + 12a_4 x^2 + \ldots\right) - 2a_2 x^2 - 6a_3 x^3 - 12a_4 x^4 - \ldots\right) +$$
$$\left(-2a_1 x - 4a_2 x^2 - 6a_3 x^3 - \ldots\right) +$$
$$\left(m^2 + m\right)a_0 + \left(m^2 + m\right)a_1 x + \left(m^2 + m\right)a_2 x^2 + \left(m^2 + m\right)a_3 x^3 + \ldots = 0$$

Then collecting terms by powers in x:

$$\left(2a_2 + \left(m^2 + m\right)a_0\right) + \left(6a_3 - 2a_1 + \left(m^2 + m\right)a_1\right)x +$$
$$\left(12a_4 - 2a_2 - 4a_2 + \left(m^2 + m\right)a_2\right)x^2 + \ldots$$

And continuing by adding the recursion pattern, we discern from the previous expression:

$$+ \ldots \left(\left(n+2\right)\left(n+1\right)a_{n+2} + \left(m^2 + m - n^2 - n\right)a_n\right)x^n = 0$$

Based on the previous expression, we solve for a_{n+2} to introduce a recursion formula:

$$a_{n+2} = \frac{-\left(m^2 + m - n^2 - n\right)a_n}{(n+2)(n+1)}$$

Which can therefore be simplified by factoring the numerator, we have a recurrence relation for the coefficients of the power series solution:

$$a_{n+2} = \frac{-(m-n)(m+n+1)}{(n+2)(n+1)}a_n$$

Now, we consider when $n = m$, $a_{n+2} = 0$, and this then leads to

$$a_{n+4}, a_{n+6}, a_{n+8} \ldots = 0$$

So, we can conclude that for *odd m* values, all *odd* coefficients a_n (for $n > m$) are zero. Also, for *even m* values, all *even* coefficients a_n (for $n > m$) are zero.

Recall that we will need two linearly independent solutions y_1, y_2 to the second order differential equation in order to form a complete solution, which can be achieved by separating terms leading to different powers (odd and even) in x.

Therefore, based on our work so far, we note that y_1 *or* y_2 will only contain a finite number of non-zero terms up to and including x^n, leading to a polynomial of degree n. Recall for the power series method, a general solution is...

$$\left[y = a_0 y_1 + a_1 y_2\right]$$

and for convenience assume $a_0 = 1;\quad a_1 = 1$

Then we have, for the first solution:

$$y_1 = 1 - \frac{m(m+1)}{(2)(1)}x^2 + \frac{(m-2)(m+2+1)}{(4)(3)}\frac{m(m+1)}{(2)}x^4 - \ldots$$

a_2 based on $n = 0$ a_4 based on $n = 2$

Moreover, for the second solution, we have

$$y_2 = 1 \cdot x + \frac{-(m-1)(m+2)}{(3)(2)}x^3 + \frac{-(m-3)(m+4)}{(5)(4)}\left(\frac{-(m-1)(m+2)}{(3)(2)}\right)x^5 - \ldots$$

These solutions can be recast as

$$\left[y_1 = 1 - \frac{m(m+1)}{2!}x^2 + \frac{(m-2)(m+3)m(m+1)}{4!}x^4 + \ldots\right]$$

and

$$\left[y_2 = x - \frac{(m-1)(m+2)}{3!}x^3 + \frac{(m-3)(m+4)(m-1)(m+2)}{5!}x^5 + \ldots\right]$$

Where we note that it can be shown that the general solution to Legendre's equation $[y = a_0 y_1 + a_1 y_2]$, convergent on $-1 < x < 1$.

Investigating what we have found a bit further,

Suppose $m = 0 \rightarrow y_1 = 1$ and other terms are zero…

$$m = 1 \rightarrow y_2 = x \text{ and other terms are zero…}$$

$$m = 2 \rightarrow y_1 = 1 - \frac{2(2+1)}{2!}x^2 + 0 \ldots \quad \text{or} \quad 1 - 3x^2$$

$$m = 3 \rightarrow y_2 = x - \frac{(3-1)(3+2)}{3!}x^3 + 0 \ldots \text{or} \quad x - \frac{5}{3}x^3$$

$$m = 4 \rightarrow y_1 = 1 - 10x^2 + \frac{35}{3}x^4$$

Legendre Polynomials are applied to be orthogonal on $[-1,1]$ so that a *normalization constant* of $\dfrac{(2m+1)}{2}$ is applied... however, because we chose for convenience

$$a_0 = 1; \quad a_1 = 1$$

the scaling of the Legendre polynomials derived here will be a bit different, where in addition to the above normalization factor, the equations for $m = \{2, 3, 4\}$ as shown should also be scaled by $\{1/2, 3/2,$ and $3/8\}$ respectively for *orthonormality* on $[-1,1]$. Verification of this orthonormality is left to the reader.

The general form for orthogonal Legendre polynomials of any order can quickly be derived using Rodrigues' formula for Legendre Polynomials, as already mentioned in Chapter 3:

$$P_n(x) = \frac{1}{2^n n!} \frac{d^n}{dx^n}(x^2 - 1)^n$$

Solution of ODEs with Variable Coefficients Using the Method of Frobenius

This is a method for obtaining a solution to an ordinary differential equation at a regular singular point applied over a deleted neighborhood. The *Method of Frobenius* (MOF) is a modified power series method. Consider the homogeneous differential equation:

$$y'' + p(x)y' + q(x)y = 0$$

having a regular singular point at $x = x_0$.

Note: For this discussion, we will assume that $x_0 = 0$; if this is not true, assume that a coordinate translation can be applied in accordance with *Axiom 1*, where $(t = x - x_0)$, is applied.

Frobenius' Theorem

If $x_0 = 0$ is a *regular singular point* of the differential equation, then it has at least one solution of the form

$$\left[y = x^{\lambda} \sum_{n=0}^{\infty} a_n x^n \right]$$

Where λ, a_n are constants, and $n = 0, 1, 2, \ldots$ valid on the interval $0 < |x| < R$. The roots λ are determined from root solving the *indicial equation*. For a second order differential equation, this will result in a *quadratic equation in* λ.

The Method of Frobenius always yields *at least one* solution of the form $\left[y_1(x) = x^{\lambda_1} \sum_{n=0}^{\infty} a_n(\lambda_1) x^n \right]$, where $\lambda_1 \geq \lambda_2$ is assumed to be the larger indicial root.

A *second solution* is required to determine the complete solution of the homogeneous equation. Therefore, a second solution can be determined from the application of *Reduction of Order*.

Given that a first solution exists, recall that a compact expression for applying reduction of order can be written as

$$\left[y_2(x) = y_1(x) \int \frac{e^{-\int p(x)dx}}{(y_1(x))^2} dx \right]$$

Applying Reduction of Order can be rigorous, and a second approach is available, depending upon the roots of the indicial equation.

Cases for a Second Solution of the Method of Frobenius

Case I: If two unique roots of the indicial equation are found, and $\lambda_1 - \lambda_2 \neq$ integer, this yields a unique second solution by applying the method of Frobenius using λ_2 in the same way it was performed using λ_1, so that

$$\left[y_2 = x^{\lambda_2} \sum_{n=0}^{\infty} a_n(\lambda_2) x^n \right]$$

Then, the general solution is $\left[y = c_1 y_1 + c_2 y_2 \right]$ applicable on the deleted neighborhood $0 < |x| < R$ of the origin, again assuming that $x_0 = 0$ is a regular singular point.

Case II:
If two repeated roots of the indicial equation are found, where $\lambda_1 = \lambda_2$, this is a repeated root which will lead to the form

$$\left[y_2 = y_1 \ln x + x^{\lambda_1} \sum_{n=0}^{\infty} b_n(\lambda_1) x^n \right]$$

where we note that this solution contains the first solution multiplied by a logarithm and summed to a second (Frobenius) series. In order to arrive at this form, we must adhere to the following procedure:

i) Recast the equation for the first MOF solution in terms of the recurrence relation for the power series solution a_n as a multivariate function in x and λ:

$$y(\lambda, x) = a_0 x^{\lambda} + a_1 x^{\lambda+1} \dots$$

ii) Then compute $\dfrac{\partial}{\partial \lambda}\left(y(\lambda, x) \right)$ and in doing so, note there will be a term requiring the computation of

$$\frac{\partial}{\partial\lambda}\left(x^{\lambda+k}\right)=\frac{\partial}{\partial\lambda}\left(\exp[(\lambda+k)\ln(x)]\right)$$

$$=\exp[(\lambda+k)\ln(x)]\ln(x)=x^{\lambda+k}\ln(x)$$

iii) Then determine the *second solution* after completing the above by substituting the repeated root into the expression, according to

$$y_2 = \frac{\partial}{\partial\lambda}\left(y(\lambda,x)\right)\Big|_{\lambda\to\lambda_2=\lambda_1}$$

This will result in the form stated above for Case II. The general solution is $[y = c_1 y_1 + c_2 y_2]$ applicable over the deleted neighborhood $0 < |x| < R$ of the origin.

Case III:

Should the two roots of the indicial equation be found, but differ by an integer value, where $\lambda_1 - \lambda_2 = N$ is a positive integer, this will lead to the following solution for y_2:

$$y_2 = h_{-1}y_1 \ln x + x^{\lambda_2}\sum_{n=0}^{\infty}h_n(\lambda_2)x^n$$

Where h_n is a constant, and the first solution y_1 appears with a new coefficient multiplied by a logarithm.

To arrive at this form, we must adhere to the following:

i) Cast the equation in terms of the a_n recurrence relation as a multivariate function in x and λ:

$$y(\lambda,x) = a_0 x^{\lambda} + a_1 x^{\lambda+1} + \dots$$

ii) Compute $\frac{\partial}{\partial\lambda}\left((\lambda-\lambda_2)y(\lambda,x)\right)$ again noting that

$$\frac{\partial}{\partial\lambda}\left(x^{\lambda+k}\right)=\frac{\partial}{\partial\lambda}\left(\exp[(\lambda+k)\ln(x)]\right)$$

$$= \exp[(\lambda + k)\ln(x)]\ln(x) = x^{\lambda+k}\ln(x)$$

iii) Determine a second solution by substituting the second root into the resulting expression according to

$$y_2 = \frac{\partial}{\partial \lambda}\big((\lambda - \lambda_2)y(\lambda, x)\big)\Big|_{\lambda \to \lambda_2}$$

This will result in the form stated above for Case III.

The general solution is $\left[y = c_1 y_1 + c_2 y_2\right]$ applicable on the deleted neighborhood $0 < |x| < R$ of the origin.

Stepwise Solution Method: Method of Frobenius

Given $y'' + p(x)y' + q(x)y = 0$

With a regular singular point identified at the origin (or translated to origin with a shift of variables as $t = (x - x_0)$), we must first assume a Frobenius Series:

$$y = x^\lambda \sum_{n=0}^{\infty} a_n x^n$$

Next, we factor terms by powers in x: $\quad x^\lambda, x^{\lambda+1}, x^{\lambda+2}$, *etc.*

Then, divide by x^λ, and *collect terms* by constants $a_0, a_1, a_2 \ldots$

Then, we identify the indicial equation (by collecting coefficients of x^0) to yield the roots $\lambda_1, \lambda_2, \ldots$ etc.

Then we attempt to identify a *recurrence relationship* (or "recurrence relation") between a_n and a_{n-1} or a_{n-2}, etc.

By Frobenius' Theorem, this procedure will provide the first solution y_1 based on the first indicial root λ_1.

To yield a second solution $y_2(x)$, either apply Reduction of Order, or identify the method (either of *Cases I, II, or III*, depending upon the roots of the indicial equation) necessary to obtain a second solution.

Example

Solve Bessel's Equation $x^2 y'' + xy' + (x^2 - v^2)y = 0$ in the neighborhood of $x_0 = 0$ for $v = 0$.

In keeping with the methods for differential equations with variable coefficients, we must analyze the equation for the point of interest. Therefore,

$$P(x) = \frac{x}{x^2} = \frac{1}{x}$$

$$Q(x) = \frac{\left(x^2 - v^2\right)}{x^2} = \left(1 - \frac{v^2}{x^2}\right)$$

Evaluating $x_0 = 0$ in these two equations indicates that it is a singular point. Therefore, as required, we then proceed to determine the type of singular point:

$$(x - 0)P(x) \rightarrow 1$$

$$(x - 0)^2 Q(x) \rightarrow \left(x^2\right)\left(1 - \frac{v^2}{x^2}\right) \rightarrow x^2 - v^2$$

We see that the point is analytic for these cases, so we then determine that $x_0 = 0$ is a *regular singular point* and the differential equation is soluble by the *Method of Frobenius*.

Then we assume the equation has the form of a Frobenius series:

$$y = x^\lambda \sum_{n=0}^{\infty} a_n x^n$$

Which results in the following derivatives for the assumed Frobenius series:

$$y = x^\lambda a_0 + a_1 x^{\lambda+1} + a_2 x^{\lambda+2} + a_{n-1} x^{\lambda+n-1} + \ldots$$

$$y' = \lambda x^{\lambda-1} a_0 + a_1(\lambda+1)x^\lambda + a_2(\lambda+2)x^{\lambda+1} + \ldots$$

$$y'' = (\lambda)(\lambda-1)a_0 x^{\lambda-2} + a_1(\lambda+1)(\lambda)x^{\lambda-1} + a_2(\lambda+2)(\lambda+1)x^\lambda + \ldots$$

Using the polynomial coefficient form of Bessel's equation, per helpful Axiom 2, then substituting into Bessel's equation:

$$x^2 y'' + xy' + (x^2 - 0^2)y = 0 \text{ in the neighborhood of } x_0 = 0.$$

$$x^2\left(a_0(\lambda)(\lambda-1)x^{\lambda-2} + a_1(\lambda+1)(\lambda)x^{\lambda-1} + a_2(\lambda+2)(\lambda+1)x^\lambda + \ldots\right)$$
$$+ x\left(a_0 \lambda x^{\lambda-1} + a_1(\lambda+1)x^\lambda + a_2(\lambda+2)x^{\lambda+1} + \ldots\right)$$
$$+ x^2\left(a_0 x^\lambda + a_1 x^{\lambda+1} + a_2 x^{\lambda+2} + a_3 x^{\lambda+3} + \ldots\right) = 0$$

Simplifying a bit:

$$\left(a_0(\lambda)(\lambda-1)x^\lambda + a_1(\lambda+1)(\lambda)x^{\lambda+1} + a_2(\lambda+2)(\lambda+1)x^{\lambda+2} + \ldots\right)$$
$$+ \left(a_0 \lambda x^\lambda + a_1(\lambda+1)x^{\lambda+1} + a_2(\lambda+2)x^{\lambda+2} + \ldots\right)$$
$$+ \left(a_0 x^{\lambda+2} + a_1 x^{\lambda+3} + a_2 x^{\lambda+4} + a_3 x^{\lambda+5} + \ldots\right) = 0$$

As outlined in the recommended procedure, we then collect terms by powers in x:

$$x^\lambda\left(a_0(\lambda^2 - \lambda) + a_0\lambda\right) + x^{\lambda+1}\left(a_1(\lambda^2 + \lambda) + a_1(\lambda+1)\right)$$
$$+ x^{\lambda+2}\left(a_2(\lambda+2)(\lambda+1) + a_2(\lambda+2) + a_0\right) + \ldots = 0$$

Which can be simplified further by collecting terms and factoring. For example, note that the expression multiplied by $x^{\lambda+2}$ can be greatly simplified, where this term is, for illustration here:

$$\left(a_2\left(\lambda^2 + 3\lambda + 2\right) + a_2\left(\lambda + 2\right) + a_0\right)$$
$$\mapsto \left(a_2(\lambda^2 + 4\lambda + 4) + a_0\right) \rightarrow \left(a_2\left(\lambda + 2\right)^2 + a_0\right)$$

This then yields a simplified equation

$$x^\lambda\left(a_0\left(\lambda^2\right)\right) + x^{\lambda+1}\left(a_1\left(\lambda + 1\right)^2\right) + x^{\lambda+2}\left(a_2\left(\lambda + 2\right)^2 + a_0\right) + \ldots = 0$$

Then, divide by x^λ, and collect terms by constants $a_0, a_1, a_2 \ldots$

Then, we identify the indicial equation (by collecting coefficients of x^0) to yield the roots $\lambda_1, \lambda_2, \ldots$ etc.

Consider the a_0 term: $a_0\left(\lambda^2\right) = 0$ and $a_0 \neq 0$

In this case, the *indicial equation* is $\left(\lambda^2\right) = 0$

Then, this leads to the a case of repeated roots $(\lambda_1 = \lambda_2 = 0)$

Furthermore, we find that from the

$$a_1(\lambda + 1)^2 = 0 \quad \text{which means that } a_1 = 0$$

This also allows us to determine that all odd coefficients must therefore be zero as well:

$$a_1 = 0 = a_3, a_5, a_7 \ldots$$

Considering the series solution equation, we can generate the general relationship for the coefficients:

$$x^{\lambda+n}\left(a_n\left(\lambda + n\right)^2 + a_{n-2}\right) = 0$$

$$a_n(\lambda + n)^2 = -a_{n-2}$$

and therefore we are able to come up with a recurrence relation

$$a_n = \left[\frac{-a_{n-2}}{(\lambda + n)^2}\right] \quad \text{for } n \geq 2$$

Therefore, to have a solution that is non-trivial (non-zero), we assume that $a_0 \neq 0$. This then enables us to determine the subsequent interdependent terms of the series, starting from a_0.

$$a_2 = \frac{-a_0}{2^2}$$

$$a_4 = \frac{-a_2}{4^2} = \frac{+\left(\dfrac{a_0}{2^2}\right)}{2^4} = \frac{a_0}{2^4 (2!)^2}$$

$$a_6 = \frac{-a_4}{6^2} = \frac{-\left(\dfrac{a_0}{2^2}\right)}{2^4}\frac{1}{6^2} = \frac{-a_0}{2^6 (3!)^2}$$

$$a_8 = \frac{-a_6}{8^2} = \frac{a_0}{2^6 (3!)^2}\frac{1}{8^2} = \frac{a_0}{2^8 (4!)^2}$$

And so on …

Placing these coefficients back into the Frobenius series, we obtain one solution to the differential equation, in accordance with Frobenius' Theorem:

$$y_1(x) = a_0\left[1 - \frac{1 \cdot x^2}{2^2 (1!)} + \frac{1 \cdot x^4}{2^4 (2!)^2} - \frac{1 \cdot x^6}{2^6 (3!)^2} + \frac{1 \cdot x^8}{2^8 (4!)^2} + \ldots\right]$$

Where we can then write the solution as a power series:

$$\left[y_1(x) = a_0\sum_{n=0}^{\infty} \frac{(-1)^n x^{2n}}{2^{2n} (n!)^2}\right]$$

Note that this is just the first solution, y_1.

To span the solution space, we need a second solution to the problem. At this point, we could apply Reduction of Order using the first solution to obtain y_2.

Alternatively, we can also obtain a second solution using the procedures for obtaining a second solution under Frobenius' Theorem, and with a repeated root, where $\lambda_2 = 0$ as Case II applies.

Recast first equation in terms of …

$$y(\lambda, x) = \left[x^\lambda a_0 - \frac{x^{\lambda+2} a_0}{(\lambda+2)^2} - \frac{x^{\lambda+4} a_2}{(\lambda+4)^2} - \frac{x^{\lambda+6} a_4}{(\lambda+6)^2} + \ldots \right]$$

Recalling that

$$a_n = \left[\frac{-a_n - 2}{-(\lambda+n)^2} \quad n \ge 2 \right]$$

$$y(\lambda, x) = a_0 x^\lambda - \frac{a_0 x^{\lambda+2}}{(\lambda+2)^2} + \frac{a_0 x^{\lambda+4}}{(\lambda+2)^2 (\lambda+4)^2} - \frac{a_0 x^{\lambda+6}}{(\lambda+2)^2 (\lambda+4)^2 (\lambda+6)^2} + \ldots$$

As a side note, consider that:

$$\frac{\partial}{\partial \lambda} x^\lambda = \frac{\partial}{\partial \lambda} e^{\ln(x^\lambda)} = \frac{\partial}{\partial \lambda} e^{\lambda \ln(x)} = e^{\lambda \ln(x)} \ln(x) = x^\lambda \ln(x)$$

So $\dfrac{\partial}{\partial \lambda} x^{\lambda+k} = x^{\lambda+k} \ln(x)$

Now, applying Case II for a second solution of the Method of Frobenius, we must take the partial derivative with respect to λ of $y(\lambda, x)$:

$$\frac{d}{d\lambda}(y_1(x,\lambda)) = a_0 x^\lambda \ln x - \frac{a_0 x^{\lambda+2} \ln x}{(\lambda+2)^2} + \frac{2a_0 x^{\lambda+2}}{(\lambda+2)^3} + \frac{a_0 x^{\lambda+4} \ln x}{(\lambda+2)^2(\lambda+4)^2}$$

$$+ \frac{-2a_0 x^{\lambda+4}(\lambda+2)^{-3}}{(\lambda+4)^2} + -2a_0 x^{\lambda+4}(\lambda+4)^{-3}(\lambda+2)^{-2} + \frac{-a_0 x^{\lambda+6} \ln x}{(\lambda+2)^2(\lambda+4)^2(\lambda+6)^2} +$$

$$\frac{-a_0 x^{\lambda+6}(-2)(\lambda+2)^{-3}}{(\lambda+4)^2(\lambda+6)^2} + \frac{-a_0 x^{\lambda+6}(-2)(\lambda+4)^{-3}}{(\lambda+2)^2(\lambda+6)^2} + \frac{-a_0 x^{\lambda+6}(-2)(\lambda+6)^{-3}}{(\lambda+2)^2(\lambda+4)^2} + \dots$$

As specified for Case II, now substitute in $\lambda_2 = 0$ in the expression

$$y_2 = \frac{\partial}{\partial \lambda}(y_1(x,\lambda))\Big|_{\lambda \to \lambda_2}$$

This results in

$$y_2 = a_0 \ln x - \frac{a_0 x^2 \ln x}{2^2} + \frac{2a_0 x^2}{2^3} + \frac{a_0 x^4 \ln x}{2^2 4^2} - \frac{2a_0 x^4}{2^3 4^2} - \frac{2a_0 x^4}{4^3 2^2} - \frac{a_0 x^6 \ln x}{2^2 4^2 6^2}$$

$$+ \frac{2a_0 x^6}{2^3 4^2 6^2} + \frac{2a_0 x^6}{2^2 4^3 6^2} + \frac{2a_0 x^6}{2^2 4^2 6^3} + \dots$$

This can be generally rewritten as

$$y_2 = a_0(\ln x)\left(1 - \frac{x^2}{2^2(1!)} + \frac{x^4}{2^4(2!)^2} + \dots\right) + a_0\left(\frac{x^2}{2^2(1!)^2}(1) - \frac{x^4}{2^4(2!)^2}\left(1 + \frac{1}{2}\right) + \dots\right)$$

Or $$\left[y_2 = y_1 \ln x + a_0\left[\frac{x^2}{2^2(1!)^2}(1) - \frac{x^4}{2^4(2!)^2}\left(1 + \frac{1}{2}\right) + \dots\right]\right]$$

And the general solution is $[y = c_1 y_1 + c_2 y_2]$.

The general solution here is often recast as

$$[y = c_1 J_0(x) + c_2 Y_0(x)]]$$

where $J_0(x)$ is an ordinary Bessel Function of the first kind of order 0, and $Y_0(x)$ is an ordinary Bessel Function of the second kind of order 0. The "zero" subscript comes from the $v = 0$ in the equation.

The general form of Bessel's Differential Equation is

$$x^2 y'' + xy' + (\lambda^2 x^2 - v^2) y = 0$$

Which has a general solution

$$[y = c_1 J_v(\lambda x) + c_2 Y_v(\lambda x)]]$$

A Modified Bessel's Differential Equation is given by

$$x^2 y'' + xy' - (\lambda^2 x^2 + v^2) y = 0$$

Which has a general solution

$$[y = c_1 I_v(\lambda x) + c_2 K_v(\lambda x)]]$$

As is evident, the Method of Frobenius is algebraically demanding, but very powerful as a means to determine a solution to certain differential equations that may contain singularities.

Bessel's differential equation is important in the solution of the neutron diffusion in cylindrical nuclear reactors, as we will demonstrate.

In neutronics, the $v = 0$ solutions arise most frequently from the form of the neutron diffusion equation, and these are demonstrated in the following plots.

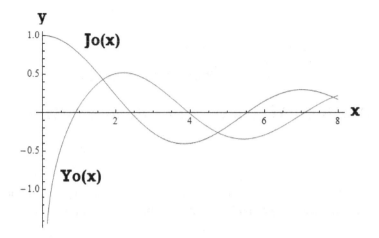

Plot of zeroth order ordinary Bessel Functions

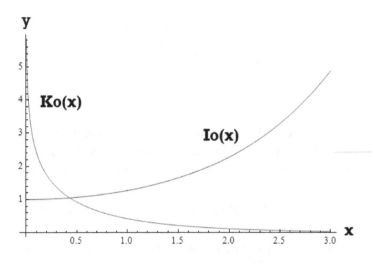

Plot of zeroth order modified Bessel Functions

Chapter 8

Vectors, Matrices, and Linear Systems

This chapter is concerns fundamental concepts in linear algebra, focusing on vectors, matrices, and their applications in linear systems.

Vectors by Example

A vector is a directed line segment from an *initial point* (A) to a *terminal point* (B), and is often denoted using the braces < > to indicate a vector quantity. This section introduces vectors with common examples.

Fundamentally, the set of all vectors with "n" real components is $\Re^n$.

Later, we will see a much wider variety of vectors and vector operations, but an understanding of ordinary vectors will remain useful when studying other vectors.

Position Vector

A *position vector* pointing to $< 3,9,2 >$ is a directed segment pointing from the origin to the terminal point at (x,y,z) coordinate $(3,9,2)$. It could also point from a different initial point than the origin. For 3-dimensional position vectors, the set of all position vectors with "3" real components is $\Re^3$.

You have already seen positions indicated by a set of coordinates, such as $(3,9,2)$. Similarly, a particle at position vector $<3,9,2>$ could make a movement along a vector $<1,3,4>$, so that its new position is $<3,9,2> + <1,3,4> = <4,12,6>$, where this *vector addition* is depicted graphically below:

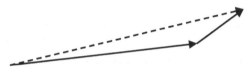

It could also make such a movement twice, so that its new position is $<3,9,2> + 2<1,3,4> = <5,15,10>$.

The operation $<3,9,2> + <1,3,4>$ is called *vector addition* because two vectors are added together.

The operation $(2 <1,3,4>)$ is called *scalar multiplication* because the vector $<1,3,4 >$ is multiplied by the scalar value 2. Note here that a position vector can be used to represent either a position or a change in position.

Velocity

Velocity is a differential change in position. If p_1 is the starting position of a particle, v is the particle's velocity, and Δt is a short amount of time (short enough to consider that v is approximately constant) then the position p_2 after the time Δt has passed is given by the following equation.

$$p_2 = p_1 + v \, \Delta t$$

Here, p_2, p_1, and v are vectors, while Δt is a scalar.

Special Coordinate Vectors

There are special unit vectors used to define the basis vectors for coordinate axes. For Cartesian coordinates, these are:

$$\hat{i} = <1,0,0>$$

$$\hat{j} = <0,1,0>$$

$$\hat{k} = <0,0,1>$$

Vector Magnitude

The magnitude, or length of any vector can be computed from the root of the sum of the squares of individual components. For a position vector $\vec{A} = <A_1, A_2, A_3>$ in three dimensional space, vector magnitude is denoted by

$$\left|\vec{A}\right| = \sqrt{A_1^2 + A_2^2 + A_3^2}$$

Dot Product

The dot product of two three dimensional vectors $\vec{A}$ and $\vec{B}$ is a scalar quantity

$$\vec{A} \cdot \vec{B} = A_1 B_1 + A_2 B_2 + A_3 B_3 = \left|\vec{A}\right|\left|\vec{B}\right| \cos\theta$$

Unit Vector

A unit vector $\hat{u}$ is a vector of magnitude 1, or "unity" in the direction of the vector from which it is formed. A unit vector in the direction of $\vec{A}$ is just the original vector divided by its magnitude:

$$\hat{u}_A = \frac{\vec{A}}{\left|\vec{A}\right|} = \frac{\vec{A}}{\sqrt{\vec{A}\cdot\vec{A}}}$$

Any vector subsequently dotted (using a dot product) with a unit vector results in the contribution of the original vector in the direction of the unit vector.

Example

The neutron flux at a point in spaced is determined to be

$$\phi(x, y, z) = \cos\left(\frac{x}{2}\right)\sin\left(\frac{y}{2}\right)\exp(-z) \qquad \frac{n}{cm^2 s}$$

Estimate the neutron current in the z-axis direction at the point $(1,1,1)$ using diffusion theory if the diffusion coefficient is $D = 2$ cm.

$$\vec{J} = -D\nabla\phi$$

$$\vec{J} = -D\left(\frac{\partial\phi}{\partial x}\hat{i} + \frac{\partial\phi}{\partial y}\hat{j} + \frac{\partial\phi}{\partial z}\hat{k}\right)$$

$$\vec{J} = -D\left(\begin{array}{l} -\frac{1}{2}\sin\left(\frac{x}{2}\right)\sin\left(\frac{y}{2}\right)\exp(-z)\hat{i} + \frac{1}{2}\cos\left(\frac{y}{2}\right)\cos\left(\frac{x}{2}\right)\exp(-z)\hat{j} + \\ -\exp(-z)\cos\left(\frac{x}{2}\right)\sin\left(\frac{y}{2}\right)\hat{k} \end{array}\right)$$

$$\vec{J}\cdot\hat{k}\Big|_p = +D\exp(-z)\cos\left(\frac{x}{2}\right)\sin\left(\frac{y}{2}\right)$$

$$= 2\exp(-1)\cos\left(\frac{1}{2}\right)\sin\left(\frac{1}{2}\right) \qquad \frac{n}{cm^2 s}$$

Orthogonality of Vectors

Vectors whose inner product is zero are orthogonal. This means that if $\vec{u}$ is orthogonal to $\vec{v}$, and both are non-zero, then $\vec{u} \cdot \vec{v} = 0$.

The inner product is sometimes written $(\vec{u}, \vec{v})$. The inner product can also be defined using functions as an integral to demonstrate orthogonality, as shown earlier. Recall from chapter 3 that the formal definition of orthogonality of functions over a fixed interval $[a,b]$ with a weighting function $w(x)$ is:

$$\int_a^b \phi_i(x)\phi_j(x)w(x)dx = \delta_{ij}$$

where δ_{ij} is Kronecker Delta Function.

A set of vectors is *orthonormal* if all vectors in the set are normal and are orthogonal to each other.

While the mathematical definitions above are useful for calculation, it is easier to think of orthogonal and normal as perpendicular and length of one.

Composing Vectors from Points

To compose a vector between any 2 points, always use the rule of "Head minus Tail" to compose the vector.

If Point P is the vector head, and Point P_1 is the vector tail, then the position vector $\vec{r}$ is composed from $\overrightarrow{P_1P}$ by subtracting P_1 from Point P.

Forming a Plane in a Medical Physics Model

You are building a high energy X-Ray treatment plan for a brain cancer patient using Monte Carlo transport. The patient's head is centered at (0,0,0). The patient treatment volume (PTV) is centered at P_1 at (1, 2, 3). The X-Ray beam source head is located at a point (-20, 2, 20). The oncologist has requested you provide a profile of the X-radiation crossing through a plane normal to the beam centered at P_1 for dose estimation and continuity of beam filtering.

Determine the vector normal to a plane where the source must point to focus on P_1. Then construct the equation of the plane normal to the X-Ray source for tallying in your Monte Carlo model.

The normal vector comes from subtracting the coordinate of the the PTV center from the coordinate of the X-Ray head.

$$\bar{N} = <-20-1, 2-2, 20-3>$$

$$\bar{N} = <-21, 0, 17>$$

Then, assume point P is a point on the plane is (x,y,z), and point P_1 is the vector tail. The position vector $\vec{r}$ is composed from $\overrightarrow{P_1 P}$ by subtracting P_1 from point P.

$$P = (x, y, z) \qquad P_1 = (1,2,3)$$

$$\overrightarrow{P_1 P} = <x-1, y-2, z-3>$$

The equation of the plane can be determined by forcing the dot product between the two vectors to be zero:

$$\overrightarrow{P_1 P} \cdot \bar{N} = 0$$

This forces the orthogonal relationship between all possible points on the plane forming a vector with P_1

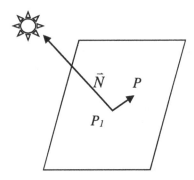

Plane depicting points normal to an X-Ray source

Therefore:

$$\overrightarrow{P_1P} \cdot \vec{N} = 0 = (x-1)(-21) + (y-2)0 + (z-3)(17) = 0$$
$$-21x + 21 + 17z - 51 = 0$$
$$\left[-21x + 17z = 30\right]$$

The equation of the plane is $-21x + 17z = 30$.

State of a System

A vector can also represent the state for an entire system, or a change to that system.

Examples include:

- a collection of particles (position and velocity for each particle)
- a sample of radioactive material (number of atoms for each relevant isotope),
- a tank containing a water-based solution (fill height, temperature, and concentration),
- a nuclear reactor (neutron density for various positions and speeds, and atomic densities for various isotopes and positions).

In each of these cases, it is important to define a format for representing the state of the system.

It is also important, but usually very easy, to define rules for adding vectors together or multiplying vectors by scalars.

Finally, for each system, physics must be applied to create rules for how the system will change with time.

Vector Spaces and Basis

Consider that n^{th} dimensional vectors exist in space as "ordered n-tuples in $\mathfrak{R}^n$ ", composed of nth dimensional real #'s.

$$\vec{a} =< a_1, a_2, a_3, a_4 ..., a_n >$$
$$\vec{b} =< b_1, b_2, b_3, b_4 ..., b_n >$$

Where $\vec{a}$ and $\vec{b}$ are non-zero, and are orthogonal if $\vec{a} \cdot \vec{b} = 0$

This then implies implies a "mathematical" concept of a space "V", where V is a vector space if...

   - <u>vector addition</u> is defined in $\mathfrak{R}^n$

   - <u>vector scalar multiplication</u> is defined in $\mathfrak{R}^n$

The axioms for *vector addition* must also hold true:

(i)   if $\vec{x}$ and $\vec{y}$ are in V, $(\vec{x} + \vec{y})$ is in V

(ii)   for <u>all</u> $\vec{x}, \vec{y}$ in V, $(\vec{x} + \vec{y} = \vec{y} + \vec{x})$

(iii)   for <u>all</u> $\vec{x}, \vec{y}, \vec{z}$ in "V, $\vec{x} + (\vec{y} + \vec{z}) = (\vec{x} + \vec{y}) + \vec{z}$

(iv)   there is a zero vector $\vec{0}$ where $\vec{x} + \vec{0} = \vec{0} + \vec{x} = \vec{x}$

(v)   for each $\vec{x}$ in V, there is a $-\vec{x}$ where $\vec{x} + (-\vec{x}) = (-\vec{x}) + \vec{x} = \vec{0}$

The axioms for *vector multiplication* must also hold true:

(vi)   for $\bar{x}$ in V, "k" is a scalar, $k\bar{x}$ in V.

(vii) $k(\bar{x}+\bar{y})=k\bar{x}+k\bar{y}$

(viii) $(k_1+k_2)\bar{x}=k_1\bar{x}+k_2\bar{x}$

(ix) $k_1(k_2\bar{x})=(k_1k_2)\bar{x}$

(x) $1\bar{x}=\bar{x}$

If the axioms for addition and multiplication for V hold true, then V is indeed a vector space. Vectors form a *Basis* for a vector space V if:

(i)   Vectors are linearly independent, which implies they are orthogonal

(ii)   the vectors *span* the vector space, e.g. $(c_1\bar{x}+c_2\bar{y})\Rightarrow$ can represent any vector $\bar{a}$ in V, and c_1 and c_2 are constants.

(iii)   linearly independent vectors that span the vector space V form a *complete set*.

(iv)   the number of vectors needed to form a complete set are said to be the *dimension* of V

Example

Consider $\vec{V}_1 =< 1,0 >$ $\vec{V}_2 =< 0,1 >$

The "dot" or inner product of these two vectors is zero:

$$\vec{V}_1\cdot\vec{V}_2=(1\cdot 0)+(0\cdot 1)=0$$

$\therefore \vec{V}_1$ is *orthogonal* to $\vec{V}_2$, and these vectors are linearly independent.

Also, the axioms for addition, multiplication hold:

$$c_1\vec{V}_1 + c_2\vec{V}_2 = \vec{a}, \text{ where } \vec{a} \text{ is any two dimensional vector}$$

$\therefore \vec{V}_1 \& \vec{V}_2$ form a complete set

Conclusion: $\{\vec{V}_1, \vec{V}_2\}$ form a *Basis* for $\Re^2$

Having discussed vector spaces, now consider the "solution space" of a differential equation:

The complete set of <u>functions</u> that form a <u>basis</u> for *all* solutions satisfying a differential equation constitute a *solution space*. We find the complete general solution, and apply specific conditions for our specific applications.

Inner Product

The inner product is defined as:

$$(\vec{x}, \vec{y}) \equiv \int_a^b x(t) * y(t) dt$$

What is an inner product space? This is a vector space with a defined inner product $(\vec{x}, \vec{y})$.

Consider

$$\vec{x}(t) = < x_1(t), x_2(t), x_3(t)...x_n(t) >$$
$$\vec{y}(t) = < y_1(t), y_2(t), y_3(t)...y_n(t) >$$

Matrices

Matrices composed of m rows by n columns can be thought of as an ordered collection of vectors.

$$\begin{bmatrix} a_{11} & a_{12} & \cdots & a_{1n} \\ a_{21} & a_{22} & \cdots & a_{2n} \\ a_{m1} & \cdots & \cdots & a_{mn} \end{bmatrix}$$

If $m = n$, the matrix is a "square matrix". If not square, it is called a *rectangular matrix*.

Matrix forms: $A\vec{x} = \vec{b}$

                A is a matrix and $\vec{x}, \vec{b}$ are column vectors

Example

$$x_1 + x_2 = 1$$
$$x_1 - x_2 = 0$$

can be directly expressed as a system of equations

$$A = \begin{bmatrix} a_{11} & a_{12} \\ a_{21} & a_{22} \end{bmatrix} = \begin{bmatrix} 1 & 1 \\ 1 & -1 \end{bmatrix} \quad \vec{x} = \begin{bmatrix} x_1 \\ x_2 \end{bmatrix} \quad \vec{b} = \begin{bmatrix} 1 \\ 0 \end{bmatrix}$$

Multiplication of matrices is performed as row by column

$$\text{For } A\vec{x} = \vec{b}$$

$$(1 \cdot x_1 + 1 \cdot x_2) = 1$$

$$(1 \cdot x_1 + -1 \cdot x_2) = 0$$

The *Rank* of a matrix is the maximum number of linearly independent equations in the matrix; this directly leads to the concept of the number of independent row vectors in the matrix. This same matrix can be written in augmented form as:

$$\begin{bmatrix} 1 & 1 & : & 1 \\ 1 & -1 & : & 0 \end{bmatrix}$$

Systems of Equations, Row Operations, and Elimination

It is often possible to solve a system of equations by converting the system into a matrix, applying "rank-preserving" operations to that matrix, and converting the matrix back into a (simpler) system of equations. The following "rank-preserving" Row Operations are used:

- *Multiply a row by a (non-zero) number.*
- *Add some multiple of one row to another row.*
- *Interchange any two rows*

Gaussian Elimination

This amounts to applying Row Operations so that the matrix is reduced to "row-echelon form". To accomplish this, the following hold true to simplify the matrix:

(i)   *The first non-zero entry in a non-zero row is a "1"*

(ii)   *"1" values appear only to the right of the row above*

(iii)   *Rows with all zeros appear at the bottom of the matrix*

Gauss–Jordan Elimination

This amounts to applying Row Operations as in Gaussian Elimination, with the added requirement that only "1" or "0" appear in the matrix, along with the solution vector in the augmented form of a matrix. This is known as "reduced row-echelon form"

The Gaussian Elimination technique will be illustrated by the following example, in which a system of three equations will be solved:

$$1\,x_1 + 4\,x_2 + 2\,x_3 = 7$$

$$8\,x_1 + 2\,x_2 + 2\,x_3 = 9$$

$$3\,x_1 + 5\,x_2 + 8\,x_3 = 7$$

The first step is to convert this system into a matrix with three rows, representing the three equations, and four columns, representing the x, y, z, and constant components of each equation.

$$\begin{bmatrix} 1 & 4 & 2 & 7 \\ 8 & 2 & 2 & 9 \\ 3 & 5 & 8 & 7 \end{bmatrix}$$

We will try to simplify the matrix, starting on the left side. First, multiply each row so that the first non-zero element is one.

$$\begin{bmatrix} 1 & 4 & 2 & 7 \\ 1 & 2/8 & 2/8 & 9/8 \\ 1 & 5/3 & 8/3 & 7/3 \end{bmatrix}$$

The leading element in the first row will be used to determine the value of x. An x-value is not needed in any other rows, so we should try to change the x-value of the other rows to zero. Perform row operations to add a multiple of the first row to each of the other rows so that the other rows have an x-value of zero. Here we subtract Row 1 from Row 2, replacing row 2; then subtract Row 1 from Row 3 and replace Row 3.

$$\begin{bmatrix} 1 & 4 & 2 & 7 \\ 0 & -15/4 & -7/4 & -47/8 \\ 0 & -7/3 & 2/3 & -14/3 \end{bmatrix}$$

As before, it is nice to have a unity value as the first non-zero element of each row. Therefore, multiply each of the rows by a constant in order to achieve this.

$$\begin{bmatrix} 1 & 4 & 2 & 7 \\ 0 & 1 & 7/15 & 47/30 \\ 0 & 1 & -2/7 & 2 \end{bmatrix}$$

The leading '1' in the second row will then be used to determine the value of x_2. We don't need this value in any of the other rows, so we should try to change the value of the other rows to zero. Add a multiple of the second row to each of the other rows so that the other rows have a x_2-value of zero. Here we subtract four times Row 2 from Row 1, replacing row 1; then subtract Row 2 from Row 3 to replace Row 3.

$$\begin{bmatrix} 1 & 0 & 2/15 & 11/15 \\ 0 & 1 & 7/15 & 47/30 \\ 0 & 0 & -79/105 & 13/30 \end{bmatrix}$$

As before, it is useful to have one as the first non-zero element of each row. Multiply each of the rows by a different number in order to achieve this.

$$\begin{bmatrix} 1 & 0 & 2/15 & 11/15 \\ 0 & 1 & 7/15 & 47/30 \\ 0 & 0 & 1 & -91/158 \end{bmatrix}$$

The leading one in the third row will be used to determine the value of z. We don't need an x_3-value in any of the other rows, so we should try to change the x_3-value of the other rows to zero. Add a multiple of the third row to each of the other rows so that the other rows have an x_3-value of zero.

$$\begin{bmatrix} 1 & 0 & 0 & 64/79 \\ 0 & 1 & 0 & 145/79 \\ 0 & 0 & 1 & -91/158 \end{bmatrix}$$

Finally, we can convert the matrix back into a system of equations. The system of equations looks much simpler now:

$$x_1 = 64/79$$

$$x_2 = 145/79$$

$$x_3 = -91/158$$

Example

Consider the following matrix:

$$A = \begin{bmatrix} 1 & 2 & 0 & -1 \\ 2 & 6 & -3 & -3 \\ 3 & 10 & -6 & -5 \end{bmatrix}$$

Determine the rank of "A" by Row Operations. With Gaussian Elimination, the following results:

$$\begin{bmatrix} 1 & 2 & 0 & -1 \\ 0 & 2 & -3 & -1 \\ 0 & 0 & 0 & 0 \end{bmatrix}$$

This is a matrix of Rank 2, where only two linearly independent vectors are present. Next, consider the following set of equations

$$x_1 + 2x_2 + 3x_3 = 2$$

$$7x_1 + 4x_2 + x_3 = 4$$

$$5x_1 + 3x_2 + x_3 = 2$$

Writing these equations in Augmented form:

$$A = \begin{bmatrix} 1 & 2 & 3 & : & 2 \\ 7 & 4 & 1 & : & 4 \\ 5 & 3 & 1 & : & 2 \end{bmatrix}$$

Row reduction of this system yields:

$$\begin{bmatrix} 1 & 0 & -1 & 0 \\ 0 & 1 & 2 & 0 \\ 0 & 0 & 0 & 1 \end{bmatrix}$$

This set of equations is *inconsistent* and *there is no solution*. The matrix is technically of Rank 2 since only two linearly independent equations are present, but as given, it is classified as an inconsistent system.

Possible outcomes of Row Reduction

(i)   *A unique solution*

In three dimensions, this is geometrically described as the single, unique point (an ordered triplet) where three unique planes cross at a single point (x, y, z).

(ii)   *Infinitely many (parametric) solutions*

In three dimensions, geometrically this is the parametric equation of a line formed where two planes cross.

Given

$$\begin{bmatrix} 1 & 3 & -2 \\ 4 & 1 & 3 \\ 2 & -5 & 7 \end{bmatrix} \begin{bmatrix} x_1 \\ x_2 \\ x_3 \end{bmatrix} = \begin{bmatrix} -7 \\ 5 \\ 19 \end{bmatrix}$$

Simplify by Gauss–Jordan elimination to yield:

$$
\begin{bmatrix} 1 & 0 & 1 \\ 0 & 1 & -1 \\ 0 & 0 & 0 \end{bmatrix}
\begin{bmatrix} x_1 \\ x_2 \\ x_3 \end{bmatrix}
=
\begin{bmatrix} 2 \\ -3 \\ 0 \end{bmatrix}
$$

This system of Rank 2 has an infinite number of solutions if cast using parametric equations. For example, let $x_3 = t$.

Then $x_2 = -3 + t$, and $x_1 = 2 - t$; this yields the parametric equation of a line.

(iii)   *No solution; inconsistent equations*

In three dimensions, geometrically, this is where there are three planes, where two of the planes are parallel to each other, creating inconsistency that prevents a unique solution to be obtained.

The Inverse and Determinants

Consider $A\vec{x} = \vec{b}$ is a linear system of equations. The Transpose of a matrix, $A^T \equiv$ switch rows and columns of matrix.

If $A = A^T$ then A is a *symmetric* matrix

If $A = -A^T$ then A is a *skew-symmetric* matrix

Consider A^{-1} the inverse of a matrix (applying only to square matrices A) yields a solution;

$$\text{If } A\vec{x} = \vec{b} \qquad \text{Then } A^{-1}\left(A\vec{x} = \vec{b}\right)$$

$$\text{Or } A^{-1}A\vec{x} = A^{-1}\vec{b} \rightarrow I\vec{x} = A^{-1}\vec{b} \Rightarrow \left[I\vec{x} = \vec{x} = A^{-1}\vec{b}\right]$$

$I \equiv$ the identity matrix, with 1's only on the diagonal elements with zeros elsewhere.

The *inverse* can be found by writing:

$$\Rightarrow \begin{bmatrix} A & : & I \end{bmatrix} \quad \text{and performing } \textit{row operations} \text{ to yield}$$

$$\Rightarrow \begin{bmatrix} I & : & A^{-1} \end{bmatrix}$$

Note that $\left[AA^{-1} = I = A^{-1}A\right]$ for a *square matrix*

Example

Find the inverse of

$$A = \begin{bmatrix} 3 & 5 \\ 1 & 4 \end{bmatrix} \rightarrow \text{a system of equations}$$

We wish to solve system 1:

$$\begin{cases} 3x_1 + 5x_2 = 10 \\ x_1 + 4x_2 = 4 \end{cases}$$

and system 2:

$$\begin{cases} 3x_1 + 5x_2 = 7 \\ x_1 + 4x_2 = 16 \end{cases}$$

Therefore, begin with $\begin{bmatrix} A & : & I \end{bmatrix}$

$$\begin{bmatrix} 3 & 5 & : & 1 & 0 \\ 1 & 4 & : & 0 & 1 \end{bmatrix} \Rightarrow \begin{bmatrix} 1 & 4 & : & 0 & 1 \\ 3 & 5 & : & 1 & 0 \end{bmatrix}$$

$$\begin{bmatrix} 1 & 4 & : & 0 & 1 \\ 0 & -7 & : & 1 & -3 \end{bmatrix} \Rightarrow \begin{bmatrix} 1 & 4 & : & 0 & 1 \\ 0 & 1 & : & -\frac{1}{7} & +\frac{3}{7} \end{bmatrix}$$

$$\begin{bmatrix} 1 & 0 & : & \frac{4}{7} & -\frac{5}{7} \\ 0 & 1 & : & -\frac{1}{7} & \frac{3}{7} \end{bmatrix}$$

$\therefore$ One determines that $A^{-1} = \begin{bmatrix} \frac{4}{7} & -\frac{5}{7} \\ -\frac{1}{7} & \frac{3}{7} \end{bmatrix}$

Note one can readily verify that $A^{-1}A = I$ to be sure all is correct using matrix multiplication.

Then to solve systems 1 and 2, and knowledge of the inverse, we can perform two matrix-vector multiplications to yield two unique solutions.

$$x = A^{-1}b = \begin{bmatrix} \frac{4}{7} & -\frac{5}{7} \\ -\frac{1}{7} & \frac{3}{7} \end{bmatrix} \begin{bmatrix} 10 \\ 4 \end{bmatrix} = \begin{bmatrix} \frac{20}{7} \\ \frac{2}{7} \end{bmatrix}$$

$$x = A^{-1}b = \begin{bmatrix} \frac{4}{7} & -\frac{5}{7} \\ -\frac{1}{7} & \frac{3}{7} \end{bmatrix} \begin{bmatrix} 7 \\ 16 \end{bmatrix} = \begin{bmatrix} -\frac{52}{7} \\ \frac{41}{7} \end{bmatrix}$$

A *determinant* is a sum of specific products composed from cofactors of a square matrix that is unique to that matrix, and is an indicator of the linear independence of the matrix.

For the 2x2 matrix *A:*

$$A = \begin{bmatrix} a_{11} & a_{12} \\ a_{21} & a_{22} \end{bmatrix}$$

The *determinant* is directly evaluated since the cofactors in a 2x2 are a_{ij} and can be directly extracted:

$$\det(A) = a_{11}a_{22} - a_{21}a_{12}$$

For the 3x3 system or n x n system, the determinant is a bit more involved. It is rendered from the sum of cofactors obtained from matrix element products formed with signed minor matrices, denoted by the sum formed from a row of terms

$$\text{where } \sum_{j=1,N} a_{ij} \left[(-1)^{i+j} M_{ij} \right].$$

This is best demonstrated by an example of a determinant computed for a 3x3 matrix:

$$\text{Find det } A \quad \text{if} \quad A = \begin{bmatrix} 1 & -3 & 3 \\ 4 & 2 & 0 \\ -2 & -7 & 5 \end{bmatrix}$$

While any row could be selected, we choose row 2 for our cofactor expansion to find the determinant to take advantage of $a_{23} = 0$, which reduces the computational effort by 1/3. Summing the terms of $a_{ij} \left[(-1)^{i+j} M_{ij} \right]$ for Row 2:

$$|A| = 4(-1)^{2+1}\begin{vmatrix} -3 & 3 \\ -7 & 5 \end{vmatrix} + 2(-1)^{2+2}\begin{vmatrix} 1 & 3 \\ -2 & 5 \end{vmatrix} + 0(-1)^{2+3}\begin{vmatrix} 1 & -3 \\ -2 & -7 \end{vmatrix}$$

$$-4(-15--21) + 2(5--6) + 0 =$$

$$(-24) + (22) = -2$$

$$\therefore \det A = |A| = -2$$

Cofactor Expansion to Find an Inverse

This method is cumbersome, but is an alternative method for finding the inverse of a matrix using a transpose of cofactors and the determinant. The method is called "cofactor expansion":

$$A^{-1} = \frac{1}{\det A}\left[(-1)^{i+j}M_{ij}\right]^{T}$$

This states that the inverse can be computed by performing the transpose of the cofactor minor matrix divided by the determinant.

Example

Determine the inverse of $A = \begin{bmatrix} 3 & 5 \\ 1 & 4 \end{bmatrix}$

For the 2x2 determinant of a matrix, this is computed from

$$\det A = (3)(4) - (1)(5) = 12 - 5 = 7$$

The inverse is then computed as

$$(-1)^{i+j}M_{ij} = \begin{bmatrix} 4 & -5 \\ -1 & 3 \end{bmatrix} \qquad A^{-1} = \frac{1}{7}\begin{bmatrix} 4 & -5 \\ -1 & 3 \end{bmatrix}$$

$$A^{-1} = \begin{bmatrix} 4/7 & -5/7 \\ -1/7 & 3/7 \end{bmatrix}$$

Additional Rules for Determinants

Note: these apply to square matrices only

- Finding det A can be made simpler through use of *row operations first.*

- If one interchanges of two rows, one must multiply (det A) by a factor of (-1).

- If two rows of a square matrix are identical, then (det A) = 0 ; A is then a singular matrix.

- If (det A) = 0, the matrix A is *singular* and has *no inverse* A^{-1}.

- $\det(cA) = C^n \det A$ for constant C and n by n matrix A.

Example

$$\text{Given } A = \begin{bmatrix} 1 & 2 & 3 \\ 7 & 4 & 1 \\ 5 & 3 & 7 \end{bmatrix} \quad \text{Find } A^{-1}$$

the inverse A^{-1} of the following matrix using the *cofactor expansion* method

$$A^{-1} = \frac{1}{\det A} \left[(-1)^{i+j} M_{ij} \right]^T$$

First, simplify with Gaussian Elimination and find the determinant:

$$\begin{bmatrix} 1 & 2 & 3 \\ 7 & 4 & 1 \\ 0 & -7 & -8 \end{bmatrix} \Rightarrow \det A = 0 + +7\begin{vmatrix} 1 & 3 \\ 7 & 1 \end{vmatrix} + -8\begin{vmatrix} 1 & 2 \\ 7 & 4 \end{vmatrix}$$

$$= 7(1-21) + -8(4-14)$$

$$= 7(-20) + -8(-10) = -60 = |A|$$

Next, we find the inverse using the cofactor expansion method. Note that we must use the original matrix for a cofactor expansion to determine the inverse A^{-1}.

$$\begin{bmatrix} -\dfrac{25}{60} & \dfrac{44}{60} & -\dfrac{1}{60} \\ \dfrac{5}{60} & +\dfrac{8}{60} & -\dfrac{7}{60} \\ \dfrac{10}{60} & -\dfrac{20}{60} & \dfrac{10}{60} \end{bmatrix} \begin{bmatrix} -\dfrac{25}{60} & \dfrac{5}{60} & \dfrac{10}{60} \\ \dfrac{44}{60} & \dfrac{8}{60} & -\dfrac{20}{60} \\ -\dfrac{1}{60} & -\dfrac{7}{60} & \dfrac{10}{60} \end{bmatrix} = \begin{bmatrix} -\dfrac{5}{12} & \dfrac{1}{12} & \dfrac{1}{6} \\ \dfrac{11}{15} & \dfrac{2}{15} & -\dfrac{1}{3} \\ -\dfrac{1}{60} & -\dfrac{7}{60} & \dfrac{1}{6} \end{bmatrix} = A^{-1}$$

Cramer's Rule

$$\text{Given} \quad A\vec{x} = \vec{b}$$

This is a matrix based solution method that enables one to solve the system by finding ratios of determinants as follows:

$$\begin{bmatrix} a_{11} & a_{12} \\ a_{21} & a_{22} \end{bmatrix} \begin{bmatrix} x_1 \\ x_2 \end{bmatrix} = \begin{bmatrix} b_1 \\ b_2 \end{bmatrix}$$

$$x_1 = \frac{\begin{vmatrix} b_1 & a_{12} \\ b_2 & a_{22} \end{vmatrix}}{\begin{vmatrix} a_{11} & a_{12} \\ a_{21} & a_{22} \end{vmatrix}} \qquad x_2 = \frac{\begin{vmatrix} a_{11} & b_1 \\ a_{21} & b_2 \end{vmatrix}}{\begin{vmatrix} a_{11} & a_{12} \\ a_{21} & a_{22} \end{vmatrix}}$$

Applies to any n by n square matrix cast in a system of equations.

Example

$$A = \begin{bmatrix} 3 & 5 \\ 1 & 4 \end{bmatrix} \qquad x = \begin{bmatrix} x_1 \\ x_2 \end{bmatrix} \qquad b = \begin{bmatrix} 10 \\ 4 \end{bmatrix}$$

$$\det A = (12 - 5) = 7$$

$$x_1 = \frac{\begin{vmatrix} 10 & 5 \\ 4 & 4 \end{vmatrix}}{7} = \frac{(40 - 20)}{7} = \begin{bmatrix} \frac{20}{7} = x_1 \end{bmatrix}$$

$$x_2 = \frac{\begin{vmatrix} 3 & 10 \\ 1 & 4 \end{vmatrix}}{7} = \frac{(12 - 10)}{7} = \begin{bmatrix} \frac{2}{7} = x_2 \end{bmatrix}$$

This can be compared with finding A^{-1} previously.

Eigenvalues and Eigenvectors

The word "eigenvalue" is German for "characteristic value" implicit in meaning aliased to some type of mathematical expression. Specifically, associated with matrices, the eigenvalues are the roots of the characteristic values of the square matrix, computed from $A\vec{K} = \lambda\vec{K}$

Which leads to

$$\det(A - \lambda I) = 0$$

$$\begin{vmatrix} (a_{11} - \lambda) & a_{12} & a_{13} \\ a_{21} & (a_{22} - \lambda) & a_{23} \\ a_{31} & a_{32} & (a_{33} - \lambda) \end{vmatrix} = 0$$

$$\det(A - \lambda I) = 0 =$$

$$\{ (a_{11} - \lambda)((a_{22} - \lambda)(a_{33} - \lambda) - a_{32}a_{23}) - a_{12}(a_{21}(a_{33} - \lambda) - a_{31}a_{23})$$

$$+ a_{13}(a_{21}a_{32} - a_{31}(a_{22} - \lambda)) \}$$

Eigenvectors are determined by returning to and solving the equation

$$A\vec{K} = \lambda\vec{K}$$

For the vector $\vec{K}$ for a given value of λ.

Row operations performed in solving this system will most often result in a matrix of reduced rank, which will then permit a parametric solution for the Eigenvector. For this reason, one of the vector components will be a free parameter, which we typically chosen to be unity (1.0). This is performed in the next example.

Example

Find all Eigenvalues and Eigenvectors of :

$$A = \begin{bmatrix} -5 & 2 \\ 2 & -2 \end{bmatrix}$$

We begin by applying the formulation

$$A\vec{K} = \lambda\vec{K} \qquad \rightarrow \qquad (A - \lambda I)\vec{K} = \vec{0}$$

And assuming the solution is non-trivial ($\vec{K} \neq \vec{0}$) we assume that

$$\det(A - \lambda I) = 0 \qquad \rightarrow \qquad (-5 - \lambda)(-2 - \lambda) - 4 = 0$$

$$(\lambda + 1)(\lambda + 6) = 0 \qquad \rightarrow \qquad \lambda_1 = -1, \ \lambda_2 = -6$$

At this point, we return to $A\vec{K} = \lambda\vec{K}$, and employ the identity matrix, whereupon we can then write $(A - \lambda I)\vec{K} = \vec{0}$. At this point only two options remain: either $\vec{K} = 0$, which is a trivial solution, or $\det(A - \lambda I) = 0$.

Again assuming the solution is non-trivial, applying each eigenvector to the case where

$$(A - \lambda_1 I)\vec{K}_1 = \vec{0} \ (A - \lambda_2 I)\vec{K}_2 = \vec{0}$$

directly leads to the eigenvectors

For $\lambda_1 = -1$, $\vec{K}_1 = \begin{bmatrix} 1 \\ 2 \end{bmatrix}$ and for $\lambda_2 = -6$, $\vec{K}_2 = \begin{bmatrix} 2 \\ -1 \end{bmatrix}$

Eigensystem Solution Method

We will investigate how to apply the eigensystem solution method using the production and decay Bateman equations as an example. To apply the eigensystem method, one simply determines eigenvalues and eigenvectors to assemble the solution.

$$\frac{dN_1}{dt} = -m_1 N_1$$

$$\frac{dN_2}{dt} = +m_1 N_1 - m_2 N_2$$

$$\frac{dN_3}{dt} = +m_2 N_2 - m_3 N_3$$

$$\begin{bmatrix} N_1' \\ N_2' \\ N_3' \end{bmatrix} = \begin{bmatrix} -m_1 & 0 & 0 \\ +m_1 & -m_2 & 0 \\ 0 & +m_2 & -m_3 \end{bmatrix} \begin{bmatrix} N_1 \\ N_2 \\ N_3 \end{bmatrix}$$

First we assume that $\vec{N} = \vec{K}e^{\lambda t}$, where $\vec{K}$ is a real eigenvector.

Then we can write

$$\vec{N}' = A\vec{N}$$
$$\vec{N}' = \lambda \vec{K}e^{\lambda t}$$

So that we can write $\lambda \vec{K}e^{\lambda t} = A\vec{K}e^{\lambda t}$ or $\lambda \vec{K} = A\vec{K}$.

Employing the identity matrix, we can write $(A - \lambda I)\vec{K} = \vec{0}$

Either $\vec{K} = 0$ is trivial solution, or $\det(A - \lambda I) = 0$

$\therefore$ assume $\vec{K} \neq$ trivial

$$\begin{bmatrix} (-m_1 - \lambda) & 0 & 0 \\ m_1 & (-m_2 - \lambda) & 0 \\ 0 & m_2 & (-m_3 - \lambda) \end{bmatrix}$$

The characteristic polynomial is $(-m_3 - \lambda)(-m_1 - \lambda)(-m_2 - \lambda) = 0$

The roots are then 3 distinct eigenvalues

$$(\lambda_1 \equiv -m_1); \quad (\lambda_2 \equiv -m_2); \quad (\lambda_3 \equiv -m_3)$$

Determine eigenvectors associated with each eigenvalue in the equations:

$$\lambda_1 = -m_1, \quad (A - \lambda I)\vec{K} = \vec{0}$$

$$\begin{bmatrix} (-m_1 + m_1) & 0 & 0 \\ m_1 & (-m_2 + m_1) & 0 \\ 0 & m_2 & (-m_3 + m_1) \end{bmatrix} \begin{bmatrix} a_1 \\ a_2 \\ a_3 \end{bmatrix} = \begin{bmatrix} 0 \\ 0 \\ 0 \end{bmatrix}$$

$$\begin{bmatrix} 0 & 0 & 0 \\ m_1 & (m_1 - m_2) & 0 \\ 0 & m_2 & (m_1 - m_3) \end{bmatrix} \begin{bmatrix} a_1 \\ a_2 \\ a_3 \end{bmatrix} = \begin{bmatrix} 0 \\ 0 \\ 0 \end{bmatrix}$$

Determining eigenvectors is analogous (often) to obtaining a parametric solution.

Choose $a_3 = 1$ since this vector system is under-defined.

Then $a_2 m_2 + (m_1 - m_3) = 0$

$$a_2 = \frac{(m_3 - m_1)}{m_2} \quad \text{and also}$$

$$a_1 m_1 + (m_1 - m_2)\frac{(m_3 - m_1)}{m_2} = 0$$

$$a_1 = \frac{(m_1 - m_2)(m_1 - m_3)}{m_1 m_2}$$

And therefore, we obtain the eigenvector associated with λ_1:

$$\vec{K}_1 = \begin{bmatrix} \dfrac{(m_1 - m_2)(m_1 - m_3)}{m_1 m_2} \\ \dfrac{(m_3 - m_1)}{m_2} \\ 1 \end{bmatrix}$$

Similarly, we can compose the eigenvectors associated with the other eigenvalues

Therefore, $\lambda_2 = -m_2$ using the second eigenvalue in system...

$$\begin{bmatrix} -m_1 + m_2 & 0 & 0 \\ m_1 & -m_2 + m_2 & 0 \\ 0 & m_2 & -m_3 + m_2 \end{bmatrix} \begin{bmatrix} b_1 \\ b_2 \\ b_3 \end{bmatrix} = \begin{bmatrix} 0 \\ 0 \\ 0 \end{bmatrix}$$

Choose $b_3 = 1$

$$b_2 m_2 + \left(m_2 - m_3\right) = 0$$
$$b_2 m_2 = m_3 - m_2$$
$$b_2 = \frac{\left(m_3 - m_2\right)}{m_2}$$

Note: $b_1 = 0$

so that $\quad \vec{K}_2 = \begin{bmatrix} 0 \\ \dfrac{\left(m_3 - m_2\right)}{m_2} \\ 1 \end{bmatrix}$

Determining the third eigenvector:

$$\begin{bmatrix} \left(-m_1 + m_3\right) & 0 & 0 \\ m_1 & -m_2 + m_3 & 0 \\ 0 & m_2 & \left(-m_3 + m_3\right) \end{bmatrix} \begin{bmatrix} d_1 \\ d_2 \\ d_3 \end{bmatrix} = \begin{bmatrix} 0 \\ 0 \\ 0 \end{bmatrix}$$

Rewriting

$$\begin{bmatrix} \left(m_3 - m_1\right) & 0 & 0 \\ m_1 & \left(m_3 - m_2\right) & 0 \\ 0 & m_2 & 0 \end{bmatrix} \begin{bmatrix} d_1 \\ d_2 \\ d_3 \end{bmatrix} = \begin{bmatrix} 0 \\ 0 \\ 0 \end{bmatrix}$$

Recall $d_3 = 1$

So,
$$\left(m_3 - m_2\right) d_2 = 0 \qquad \therefore d_2 = 0$$
$$\left(m_3 - m_1\right) d_1 = 0 \qquad \therefore d_1 = 0$$

$$\vec{K}_3 = \begin{bmatrix} 0 \\ 0 \\ 1 \end{bmatrix}$$

So that this is the third eigenvector.

By superposition in linear systems, we may write the solution

$$\left\{ \vec{N} = C_1 \vec{K}_1 e^{\lambda_1 t} + C_2 \vec{K}_2 e^{\lambda_2 t} + C_3 \vec{K}_3 e^{\lambda_3 t} \right\}$$

$$\vec{N} = C_1 \begin{bmatrix} \dfrac{(m_1 - m_2)(m_1 - m_3)}{m_1 m_2} \\ \dfrac{(m_3 - m_1)}{m_2} \\ 1 \end{bmatrix} e^{-m_1 t} + C_2 \begin{bmatrix} 0 \\ \dfrac{(m_3 - m_2)}{m_2} \\ 1 \end{bmatrix} e^{-m_2 t} + C_3 \begin{bmatrix} 0 \\ 0 \\ 1 \end{bmatrix} e^{-m_3 t} = \begin{bmatrix} N_1 \\ N_2 \\ N_3 \end{bmatrix}$$

We can then apply an initial condition

$$\vec{N}(0) = \begin{bmatrix} N_{10} \\ 0 \\ 0 \end{bmatrix} \quad \text{So…}$$

$$\begin{bmatrix} N_{10} \\ 0 \\ 0 \end{bmatrix} = C_1 \begin{bmatrix} \dfrac{(m_1 - m_2)(m_1 - m_3)}{m_1 m_2} \\ \dfrac{(m_3 - m_1)}{m_2} \\ 1 \end{bmatrix} + C_2 \begin{bmatrix} 0 \\ \dfrac{(m_3 - m_2)}{m_2} \\ 1 \end{bmatrix} + C_3 \begin{bmatrix} 0 \\ 0 \\ 1 \end{bmatrix}$$

$$\therefore N_{10} = C_1 \frac{(m_1 - m_2)(m_1 - m_3)}{m_1 m_2} + 0 + 0$$

$$C_1 = \left[\frac{N_{10} m_1 m_2}{(m_1 - m_2)(m_1 - m_3)} \right]$$

Then

$$0 = \frac{N_{10} m_1 m_2 (-1)}{(m_1 - m_2)(m_1 - m_3)} \frac{(-m_3 + m_1)}{m_2} + C_2 \frac{(m_3 - m_2)}{m_2}$$

$$C_2 \frac{(m_3 - m_2)}{m_2} = + \frac{N_{10} m_1}{(m_1 - m_2)}$$

$$C_2 = \left[\frac{N_{10} m_1 m_2}{(m_1 - m_2)(m_3 - m_2)} \right]$$

$$0 = \frac{N_{10} m_1 m_2}{(m_1 - m_2)(m_1 - m_3)} + \frac{N_{10} m_1 m_2}{(m_1 - m_2)(m_3 - m_2)} + C_3$$

$$C_3 = -N_{10} m_1 m_2 \left(\frac{1}{(m_1 - m_2)(m_1 - m_3)} + \frac{1}{(m_1 - m_2)(m_3 - m_2)} \right)$$

Then

$$C_3 = -N_{10} m_1 m_2 \left(\frac{(m_3 - m_2) + (m_1 - m_3)}{(m_1 - m_2)(m_1 - m_3)(m_3 - m_2)} \right)$$

$$C_3 = -N_{10} m_1 m_2 \left(\frac{(m_1 - m_2)}{(m_1 - m_2)(m_1 - m_3)(m_3 - m_2)} \right)$$

$$C_3 = \left[\frac{N_{10} m_1 m_2}{(m_1 - m_3)(m_2 - m_3)} \right]$$

The complete solution is cast as

$$
\bar{N} = \begin{bmatrix} N_1 \\ N_2 \\ N_3 \end{bmatrix} = \frac{N_{10}m_1m_2}{(m_1 - m_2)(m_1 - m_3)} \begin{bmatrix} \dfrac{(m_1 - m_2)(m_1 - m_3)}{m_1m_2} \\ \dfrac{(m_3 - m_1)}{m_2} \\ 1 \end{bmatrix} e^{-m_1 t}
$$

$$
+ \frac{N_{10}m_1m_2}{(m_1 - m_2)(m_3 - m_2)} \begin{bmatrix} 0 \\ \dfrac{(m_3 - m_2)}{m_2} \\ 1 \end{bmatrix} e^{-m_2 t} + \frac{N_{10}m_1m_2}{(m_1 - m_3)(m_2 - m_3)} \begin{bmatrix} 0 \\ 0 \\ 1 \end{bmatrix} e^{-m_3 t}
$$

$$
\bar{N} = \begin{bmatrix} N_1 \\ N_2 \\ N_3 \end{bmatrix} = \begin{bmatrix} N_{10}e^{-m_1 t} \\ -\dfrac{N_{10}m_1 e^{-m_1 t}}{(m_1 - m_2)} \\ \dfrac{N_{10}m_1m_2 e^{-m_1 t}}{(m_1 - m_2)(m_1 - m_3)} \end{bmatrix} + \begin{bmatrix} 0 \\ \dfrac{N_{10}m_1 e^{-m_2 t}}{(m_1 - m_2)} \\ \dfrac{N_{10}m_1m_2 e^{-m_2 t}}{(m_1 - m_2)(m_3 - m_2)} \end{bmatrix} + \begin{bmatrix} 0 \\ 0 \\ \dfrac{N_{10}m_1m_2 e^{-m_3 t}}{(m_1 - m_3)(m_2 - m_3)} \end{bmatrix}
$$

Cryptography and Matrices

People involved in making secure communications codes (our nation's *National Security Agency*) are experts at generating prime numbers (used to make unique keys and codes) and performing matrix arithmetic.

Consider the following 7 x 3 encoded message E_m matrix:

$$E_m = \begin{pmatrix} 240 & 281 & 214 \\ 300 & 361 & 314 \\ 292 & 208 & 220 \\ 177 & 283 & 245 \\ 188 & 177 & 250 \\ 260 & 186 & 304 \\ 194 & 310 & 186 \end{pmatrix}$$

Now consider a pre-determined *Code Matrix*:

Code Matrix =

$$\begin{bmatrix} 12 & 17 & 8 & 21 & 27 & 19 & 3 & 25 & 20 & 11 & 6 & 2 & 16 & 4 & 18 & 15 & 22 & 9 & 14 & 5 & 26 & 7 & 23 & 1 \\ A & B & C & D & E & F & G & H & I & J & K & L & M & N & O & P & Q & R & S & T & U & V & W & X \end{bmatrix}$$

$$\begin{bmatrix} 13 & 10 & 24 \\ Y & Z & - \end{bmatrix}$$

Now consider a "crypto-key" matrix that can change week to week...

This week the Crytographic Code Matrix is $K = \begin{bmatrix} 4 & 9 & 2 \\ 3 & 5 & 7 \\ 8 & 1 & 6 \end{bmatrix}$

In a matrix product for this case, we have $7 \times 3 \cdot 3 \times 3 \rightarrow 7 \times 3$

$$M \cdot K = E_m \qquad M \cdot K \cdot K^{-1} = E_m K^{-1} \qquad M = E_m K^{-1}$$

Finding the inverse to the key matrix, we obtain

$$
K^{-1} = \begin{pmatrix}
\dfrac{23}{360} & -\dfrac{13}{90} & \dfrac{53}{360} \\[2ex]
\dfrac{19}{180} & \dfrac{1}{45} & -\dfrac{11}{180} \\[2ex]
-\dfrac{37}{360} & \dfrac{17}{90} & -\dfrac{7}{360}
\end{pmatrix}
$$

We then apply this so that $M = E_m K^{-1}$ which yields the message

$$
M = \begin{pmatrix}
23 & 12 & 14 \\
25 & 24 & 16 \\
18 & 4 & 26 \\
16 & 27 & 4 \\
5 & 24 & 12 \\
5 & 24 & 21 \\
26 & 14 & 6
\end{pmatrix}
$$

Which translates using the Code Matrix to the decoded message:

"wash-monument-at-dusk"

Chapter 9

Gram–Schmidt Orthogonalization and Fourier Series

This chapter concerns fundamental concepts series expansions of functions as a preamble to the solution of partial differential equations. Mastery of these topics are important in order to be well prepared to solve partial differential equations.

Gram–Schmidt Orthogonalization

The Gram–Schmidt Procedure is one that is used to derive an orthogonal set of basis vectors $\{\, \vec{u}_1, \vec{u}_2 \,\}$ using two vectors $\{\, \vec{v}_1, \vec{v}_2 \,\}$.

The solution steps for this procedure for n-dimensional systems is as follows:

$$\text{Let } \vec{u}_1 = \vec{v}_1$$

$$\text{Determine } \vec{u}_2 = \vec{v}_2 - \left(\vec{v}_2 \cdot \frac{\vec{u}_1}{|\vec{u}_1|} \right) \frac{\vec{u}_1}{|\vec{u}_1|}$$

$$\text{Determine } \vec{u}_3 = \vec{v}_3 - \left(\vec{v}_3 \cdot \frac{\vec{u}_1}{|\vec{u}_1|} \right) \frac{\vec{u}_1}{|\vec{u}_1|} - \left(\vec{v}_3 \cdot \frac{\vec{u}_2}{|\vec{u}_2|} \right) \frac{\vec{u}_2}{|\vec{u}_2|}$$

Continue the procedure to form an *n*-dimensional orthogonal set of vectors. Note that this procedure subtracts the contribution of $\vec{v}_2$ that is in the direction of $\vec{u}_1$, removing any dependence of $\vec{u}_2$ on $\vec{u}_1$. The orthogonality of the two vectors can be demonstrated by taking the dot product $\vec{u}_1 \cdot \vec{u}_2$ and demonstrating that it is indeed zero. To demonstrate, consider two vectors in space.

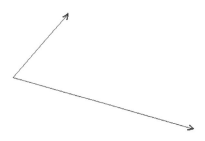

Two vectors in space

We can demonstrate this readily with *Mathematica*, where we define the two vectors $\vec{v}_1, \vec{v}_2$ and apply the Gram–Schmidt procedures:

```
In[1]:= v1 = {2, -1};
        v2 = {1, 1};

In[3]:= u1 = v1
        mag[v_] = √v.v;
        u2 = v2 - ( v2 . u1 / mag[u1] ) u1 / mag[u1]

Out[3]= {2, -1}

Out[5]= { 3/5, 6/5 }|

In[6]:= u1.u2

Out[6]= 0
```

Example

Given vectors

$$\vec{v}_1 = 2\hat{i} - 1\hat{j} + 0\hat{k}, \quad \vec{v}_2 = 1\hat{i} + 1\hat{j} + 3\hat{k}, \quad \vec{v}_3 = 1\hat{i} + 0\hat{j} + 5\hat{k}$$

Using Gram–Schmidt Procedures, derive orthogonal basis vectors and demonstrate that the basis vectors are mutually orthogonal in $\Re^3$. How can we make these orthonormal?

Again, through the use of *Mathematica*, we can solve this problem.

```
In[1]:= v1 = {2, -1, 0};
        v2 = {1, 1, 3};
        v3 = {1, 0, 5};

In[4]:= u1 = v1
        mag[v_] = √v.v ;
        u2 = v2 - ( v2 . u1 / mag[u1] )  u1 / mag[u1]

        u3 = v3 - ( v3 . u1 / mag[u1] )  u1 / mag[u1] - ( v3 . u2 / mag[u2] )  u2 / mag[u2]

Out[4]= {2, -1, 0}

Out[6]= { 3/5 , 6/5 , 3 }

Out[7]= { - 2/3 , - 4/3 , 2/3 }

In[8]:= u1.u2

Out[8]= 0

In[9]:= u1.u3

Out[9]= 0

In[10]:= u2.u3

Out[10]= 0
```

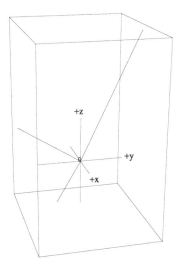

Plot of resulting orthogonal basis vectors

The orthogonal basis vectors that result,

$$\vec{u}_1 = 2\hat{i} - 1\hat{j} + 0\hat{k}$$

$$\vec{u}_2 = \frac{3}{5}\hat{i} + \frac{6}{5}\hat{j} + 3\hat{k}$$

$$\vec{u}_3 = \frac{-2}{3}\hat{i} + \frac{-4}{3}\hat{j} + \frac{2}{3}\hat{k}$$

These can be made mutually *orthonormal* by simply converting them to unit vectors.

Weierstrass's Theorem

Weierstrass's Theorem states that we can expand an *arbitrary continuous function* on a finite closed interval [A, B] using a set of powers in x to yield uniform convergence over that interval (convergence in the mean). This means that we can *represent the function with powers in x as precisely as we want to*. Note this is similar to the expansion of a function using Fourier series, to be covered later.

Simply put, Weierstrass's Theorem states that we can *expand any arbitrary continuous function* on a finite closed interval [A, B] using a set of orthogonal functions.

Gram–Schmidt procedures can be used on functions along with the orthogonality property successively. Recall the orthogonality property is:

$$\delta_{m,n} = \int_{A}^{B} P_n(x)P_m(x)w(x)dx$$

where $w(x)$ is a weighting function determined for interval [A, B]; the orthogonality property is used to successively yield appropriate constants in the polynomials.

Therefore, we wish to construct polynomials $P_0, P_1, P_2 \ldots$ and these are to be orthogonal functions on $[A, B]$

Note: one may assume $w(x) = 1$ for polynomials in x. For other functions, this may not hold true. For example, for Bessel functions, which are also orthogonal functions, $w(x) = x$ for orthogonality over the interval.

To build polynomials in x, first set $P_0(x) = 1$.

Then, using a Gram–Schmidt protocol, we assume that we can then construct an orthogonal polynomial by subtracting off the contribution of the first function:

$$P_1(x) = x - \alpha_1 P_0(x)$$

$$\text{Or} \quad P_1(x) = x - \alpha_1$$

Then the equation $\int_A^B P_0(x)P_1(x)dx = 0$ then permits us to solve for the constant α_1 that will establish orthogonality between $P_0(x)$ and $P_1(x)$.

Now, we continue the process. Again by Gram–Schmidt protocol, we next assume that we can then construct a third orthogonal polynomial by subtracting off the contributions of the first two functions:

$$P_2(x) = x^2 - \beta_2 P_1(x) - \alpha_2 P_0(x)$$

Constrained by the following orthogonality conditions:

$$\int_A^B P_2(x)P_1(x)dx = 0 \quad and \quad \int_A^B P_2(x)P_0(x)dx = 0$$

we can then solve for α_2 and β_2. This enables us to solve for the constants α_2 and β_2 that will establish orthogonality between both $P_0(x)$ and $P_2(x)$, as well as between $P_1(x)$ and $P_2(x)$.

The process can be continued to a degree as high as required for the polynomials $P_0, P_1, P_2 \ldots P_n$ to be constructed as mutually orthogonal polynomial functions on $[A, B]$.

Approximating Functions

Once we have established the orthogonal polynomials on $[A, B]$, we can then use them to approximate any analytic, well-behaved function $f(x)$ using:

$$f(x) = \sum_{n=0}^{\infty} a_n P_n(x) = a_0 P_0 + a_1 P_1 + a_2 P_2 + ... \quad \text{on } [A, B]$$

Where the constant coefficients a_n are called the moments of the orthogonal functions.

To obtain the zeroth moment a_0 ...,

- Multiply both sides of the above expansion equation by P_0
- Integrate over domain $[A, B]$

$$\int_A^B f(x) P_0(x) dx = \int_A^B (a_0 P_0^2 + a_1 P_0 P_1 + a_2 P_0 P_2 + ...) dx$$

$$\downarrow \qquad \downarrow$$
$$0 \qquad 0$$

all other terms are zero by orthogonality

Then, we can directly solve for $\left[a_0 \Rightarrow \dfrac{\int_A^B f(x) P_0 dx}{\int_A^B P_0^2 dx} \right]$ as the only

surviving term

So that

$$a_0 = \left[c_0 \int_A^B f(x)P_0 dx \right] \qquad \text{where} \qquad c_0 \equiv \left(\int_A^B (P_0(x))^2 dx \right)^{-1}$$

The nth moment of the function $f(x)$ is defined by:

$$a_n = \left[c_n \int_A^B f(x)P_n dx \right] \qquad \text{where} \qquad c_n \equiv \left(\int_A^B (P_n(x))^2 dx \right)^{-1}$$

A similar procedure is used to solve for the other coefficients.

Example

On the interval $[0, \Delta x]$ using the Gram-Schmidt procedures with Weierstrass's Theorem, it can be shown that:

$$P_0(x) = 1 \qquad\qquad\qquad c_0 = \frac{1}{\Delta x}$$

$$P_1(x) = x - \frac{\Delta x}{2} \qquad\qquad\qquad c_1 = \frac{12}{\Delta x^3}$$

$$P_2(x) = x^2 - x\Delta x + \frac{\Delta x^2}{6} \qquad\qquad\qquad c_2 = \frac{180}{\Delta x^5}$$

As an example of their application, we can use these orthogonal functions to approximate $f(x) = \cos\left(\dfrac{x}{\Delta x}\right)$

To demonstrate this, we again enlist the help of *Mathematica*, where 'p0', 'p1', and 'p2' have already been defined in *Mathematica* as given above:

In[52]:= **f[x_] := Cos[x / Δx] ;**

In[53]:= **a0 = c0 $\int_0^{Δx}$ f[x] p0 dx**

Out[53]= Sin[1]

In[54]:= **a1 = c1 $\int_0^{Δx}$ f[x] p1 dx**

Out[54]= $\dfrac{6\,(-2 + 2\,\text{Cos}[1] + \text{Sin}[1])}{Δx}$

In[55]:= **a2 = c2 $\int_0^{Δx}$ f[x] p2 dx**

Out[55]= $\dfrac{30\,(6 + 6\,\text{Cos}[1] - 11\,\text{Sin}[1])}{Δx^2}$

In[56]:= **fest[x_] := a0 p0 + a1 p1 + a2 p2 // Expand**

In[57]:= **fest[x] /. Δx -> 2 // N**

Out[57]= $1.00341 - 0.0182682\,x - 0.107752\,x^2$

Plotting this polynomial along with the original function, for $Δx = 2$ is:

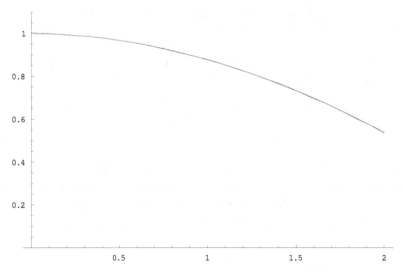

Plot of $f_{est}(x) = 1.00341 - 0.0182682x - 0.107752x^2$ and $f(x) = \cos(x/2)$

NUCLEAR APPLICATION: *Representation of Scattering Cross Sections*

Nuclear scattering cross sections typically are represented using Legendre Polynomials in order to standardize how they can be read into radiation transport codes. This is natural, since the cross sections are based on direction cosines $[0, \pi]$ that range from $[-1,1]$ that are naturally mapped by the Legendre Polynomials orthogonal over that same interval, where $\mu = Cos\,\theta$.

Example

Consider that a certain nuclide exhibits the following scattering cross section as a function of neutron scatter angle (direction cosine), in barns:

$$\sigma_s(\mu) = 10(1 + Sin(\mu)) \text{ barns}$$

Let us assume that we choose to truncate this at a second order function of the first three Legendre Polynomials as represented by $f(\mu)$:

$$f(\mu) = \left(\sigma_{s0}P_0(\mu) + \sigma_{s1}P_1(\mu) + \sigma_{s2}P_2(\mu) \right)$$

Therefore, we recognize that by truncating the order to 2, we can only approximate the cross section so that:

$$\sigma_s(\mu) \approx f(\mu)$$

The next step required is of course to solve for the three "scattering moments" $\sigma_{s0}, \sigma_{s1}, \sigma_{s2}$ by applying orthogonality, where we multiply both sides of the above equation by $P_0(\mu)$, $P_1(\mu)$, and $P_2(\mu)$, respectively, and each time integrating both sides from $[-1,1]$.

This produces the following results:

$$\sigma_{S0} = 10; \quad \sigma_{S1} = -30\left(Cos\left[1\right] - Sin\left[1\right]\right); \quad \sigma_{S2} = 0$$

and results in: $\sigma_S(\mu) \approx f(\mu) = 10 - 30\mu\left(Cos[1] - Sin[1]\right)$ barns

We can plot the original function as well as the truncated Legendre moment based linear equation; this fit is quite good, and demonstrates the power of the Legendre moments.

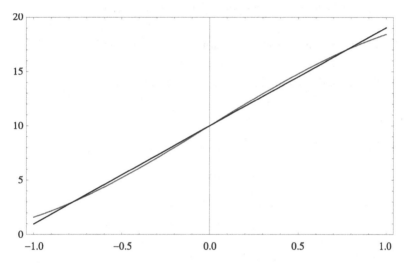

Representation of original cross section and Legendre moment equation representing $10(1 + Sin(\mu))$

As an exercise, determine the 0th, 1st, and 2nd moments for the following scattering cross section:

$$\sigma_S(\mu) = 10(1 + \mu^4) \text{ barns}$$

Fourier Series

A Fourier Series is a series expansion of a periodic function. A function f (x) with a *period* (T) of 2π can be represented by the series:

$$\left[f(x) = a_0 + \sum_{n=1}^{\infty} a_n \cos(nx) + b_n \sin(nx) \right]$$

Fourier first discovered this series in his comprehensive study of heat conduction, and it will be seen that these functions arise in *boundary conditions of heat conduction and diffusion problems.*

Note that $f(x)$ is a periodic function, and $\cos(nx)$ and $\sin(nx)$ must be *orthogonal* in order to solve for the *Fourier coefficients* a_n and b_n. These coefficients are determined by the so-called Euler formulas.

The "Euler formulas" include

$$\begin{cases} a_0 = \frac{1}{2\pi} \int_{-\pi}^{\pi} f(x)dx \\[2ex] a_n = \frac{1}{\pi} \int_{-\pi}^{\pi} f(x)\cos(nx)dx \quad n = 1,2,3... \\[2ex] b_n = \frac{1}{\pi} \int_{-\pi}^{\pi} f(x)\sin(nx)dx \quad n = 1,2,3... \end{cases}$$

In the Euler formulas, the "$\frac{1}{\pi}$" term comes from the application of orthogonality.

Both sides of the Fourier Series equation are multiplied by either the cosine or sine terms, and one is left to solve for the coefficients a_n and b_n:

$$\left(\int\limits_{-\pi}^{\pi} (\cos(nx))^2 \, dx \right)^{-1} = \frac{1}{\pi} = \left(\int\limits_{-\pi}^{\pi} (\sin(nx))^2 \, dx \right)^{-1}$$

Moreover, we note that $f(x)$ is periodic over $(-\pi, \pi)$, and finite and bounded over $(-\pi, \pi)$ with limits that exist, where left and right hand limits are

$$\left[f(x^+) = \lim_{\varepsilon \to 0} f(x+\varepsilon) \right] \quad \left[f(x^-) = \lim_{\varepsilon \to 0} f(x-\varepsilon) \right]$$

So that the series converges to $\left[\frac{1}{2} \left(f(x^+) + f(x^-) \right) \right]$, an average of left and right limits.

If $f(x)$ is *discontinuous* at "x", then the series converges to the *mean* of the two limits.

Example

$$f(x) = \begin{cases} -K & if & -\pi < x < 0 \\ +K & if & 0 < x < \pi \end{cases} \quad \begin{array}{l} with \;\; f(x+2\pi) = f(x) \\ \quad\quad (periodic) \end{array}$$

Note that if one is given a piecewise continuous function, you must perform piecewise integration.

$$a_0 = \frac{1}{2\pi}\left(\int_{-\pi}^{0} -K dx + \int_{0}^{\pi} K dx\right) = \frac{1}{2\pi}\left(-Kx\Big]_{-\pi}^{0} + Kx\Big]_{0}^{\pi}\right)$$

$$a_0 = \frac{1}{2\pi}\left(-K(0++\pi)+ K(\pi-0)\right) = 0$$

$$a_n = \frac{1}{\pi}\left[\int_{-\pi}^{0} -K\cos(nx)dx + \int_{0}^{\pi} K\cos(nx)dx\right]$$

$$a_n = \frac{1}{\pi}\left[\frac{-K}{n}\sin(nx)\right]_{-\pi}^{0} + \frac{1}{\pi}\left[\frac{K}{n}\sin(nx)\right]_{0}^{\pi} = 0$$

$$\therefore a_n = 0$$

$$b_n = \frac{1}{\pi}\left[\int_{-\pi}^{0} -K\sin(nx)dx + \int_{0}^{\pi} K\sin(nx)dx\right]$$

$$b_n = \frac{1}{\pi}\left[\frac{+K}{n}\cos(nx)\right]_{-\pi}^{0} + \frac{1}{\pi}\left[\frac{-K}{n}\cos(nx)\right]_{0}^{\pi}$$

$$b_n = \frac{1}{\pi}\left[\frac{K}{n}\cos(n\cdot0) - \frac{K}{n}(-n\pi)\right] + \frac{1}{\pi}\left[\frac{-K}{n}\cos(n\pi) + \frac{K}{n}\cos(n\cdot0)\right]$$

$$b_n = \frac{1}{\pi}\left[\frac{K}{n} - \frac{K}{n}\cos(n\pi) - \frac{K}{n}\cos(n\pi) + \frac{K}{n}\right]$$

$$\text{or}\quad b_n = \frac{2K}{n\pi}(1-\cos(n\pi))$$

$$n=1\quad \cos(\pi)\to -1$$
$$n=2\quad \cos(2\pi)\to 1$$

$$\left[b_n = \frac{2K}{n\pi}(1-(-1)^n)\right]$$

In summary,

$$a_0 = 0 \quad a_n = 0$$

$$b_n = \frac{2K}{n\pi}\left(1-(-1)^n\right)$$

Then

$$f(x) = \sum_{n=1}^{\infty} \frac{2K}{n\pi}\left(1-(-1)^n\right)\sin(nx)$$

$$f(x) = \frac{2K}{\pi}\left(2\sin x\right) + \frac{2K}{2\pi}0\sin(2x) + \frac{2K}{3\pi}2\sin(3x) + ...$$

$$f(x) = \frac{4K}{\pi}\sin x + \frac{4K}{3\pi}\sin(3x) + ...$$

so

$$\left[f(x) = \sum_{n=1}^{\infty} \frac{4K}{n\pi}\sin(nx) \right]$$

This is the Fourier series for the square wave, and we note it is an "odd" function (symmetrically opposing signs across the axis). If we plot this Fourier Series truncated to $n = 7$, we obtain the following:

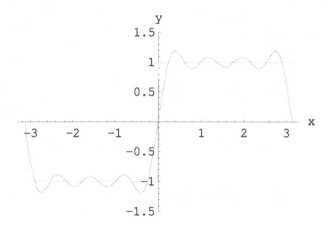

Plot of step function Fourier Series truncated at $n = 7$

Even vs. Odd Functions

Recall *cosine* is an *even* function, and the *sine* is an *odd* function.

- With $f(x)$ *odd*, all cosine terms $(a_n's) \rightarrow 0$.
    - This leads to a Fourier *sine* series.

- With $f(x)$ *even*, all sine terms $(b_n's) \rightarrow 0$.
    - This leads to a Fourier *cosine* series.

The Gibbs Phenomenon

J. Willard Gibbs (1839–1903) noted that in general, with an increasing number of terms, the Fourier Series better reproduces $f(x)$, even in the neighborhood discontinuities of $f(x)$...

However, at the specific points of discontinuity the series presents notable deviations from $f(x)$ regardless of how many terms are included in the series.

Hence, this has been named *"The Gibbs Phenomenon."*

As an example of the Gibbs phenomenon, we plot the square wave function for increasing orders of n = 11 and n = 15. The "overshoot" of the Gibbs is clearly indicated.

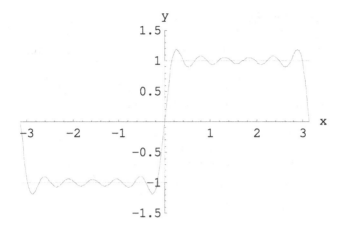

Plot of step function Fourier Series truncated at $n = 11$

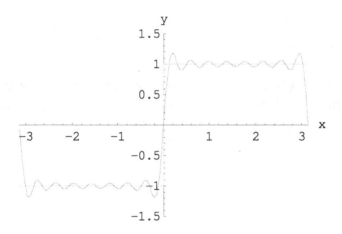

Plot of step function Fourier Series truncated at $n = 15$

Sometimes, special handling may be used to treat the Gibbs regions near discontinuities; these are often termed as a "jump" conditions.

Functions of any Period

Functions of any period can be used by a contraction ("stretching") of $p = 2\pi$ to $p = 2L$

Then a function of period $2L = p$ has a trigonometric series:

$$
\begin{bmatrix}
f(x) = a_0 + \sum_{n=1}^{\infty} \left(a_n \cos\left(\frac{n\pi x}{L}\right) + b_n \sin\left(\frac{n\pi x}{L}\right) \right) \\
a_0 = \frac{1}{2L} \int_{-L}^{L} f(x)\,dx \\
a_n = \frac{1}{L} \int_{-L}^{+L} f(x) \cos\left(\frac{n\pi x}{L}\right) dx \qquad b_n = \frac{1}{L} \int_{-L}^{L} f(x)\sin\left(\frac{n\pi x}{L}\right) dx
\end{bmatrix}
$$

$$n = 1,2,3\ldots \qquad\qquad\qquad n = 1,2,3\ldots$$

Overall, the application of orthogonality is essential in solving key engineering partial differential equations, and we will apply numerous concepts from this chapter in the work ahead.

Chapter 10

Applied Solution Methods—Part 1

Up to this point, we have covered a number of topics that form a mosaic of tools and methods useful in solving applied differential equations problems. There remain some critical applied engineering concepts that are essential to completing the suite of tools desirable for an engineer to become highly effective at solving differential equations, both ordinary and partial. This chapter, titled "Applied Solution Methods—Part 1" covers those remaining topics. Following this chapter is "Applied Solution Methods—Part 2 PDEs and Heat Transfer" which focuses on problem solving methods based on the heat equation and the neutron diffusion equation.

Coordinate Systems, Revisited

We briefly presented the most commonly used coordinate systems in Chapter 3. Ultimately, all coordinate systems stem from Cartesian coordinates, the familiar x, y, z gridpoint in 3-Dimensional space.

In addition, all vector operations for functions must also be properly derived in any other coordinate system, and vector operations imply the use of the "Del" vector operator. Recall that in Cartesian Coordinates, the Del operator is

$$\nabla = \left\langle \frac{\partial}{\partial x}, \frac{\partial}{\partial y}, \frac{\partial}{\partial z} \right\rangle = \frac{\partial}{\partial x}\hat{i} + \frac{\partial}{\partial y}\hat{j} + \frac{\partial}{\partial z}\hat{k}$$

Note that by itself, this operator does not mean anything; it is merely an operator that must operate on something; ordinarily, the object upon which it operates defines the outcome. The Laplacian, ∇^2 in Cartesian Coordinates, is computed from

$$\nabla^2 = \nabla \cdot \nabla = \frac{\partial^2}{\partial x^2} + \frac{\partial^2}{\partial y^2} + \frac{\partial^2}{\partial z^2}$$

We must note here that the Laplacian is a scalar quantity. Also, the relative ease with which the Laplacian is computed in Cartesian coordinates stems from the fact that the relative scale factors along the different axes, x, y, z are each unity:

> *a relative change of one single unit measure along each dimension x, y, or z has exactly the same measure relative to the other coordinate axes*

This is not necessarily true for curvilinear coordinate systems, and therefore the Del operator computed for curvilinear systems must be derived in a manner that accounts for the relative scaling factors relating each axis; these may be derived directly using Cartesian coordinates. If this sounds confusing, it will be made clearer through the use of an example. Before this, we will review some basic Cartesian operations using the Del Operator. While we have used them previously, we will more formally define vector operations here, as they form what is needed for PDE applications later.

The Gradient

$\vec{F} = \nabla f$, the gradient of a scalar function f

Physically, in two dimensions (x, y), or "2-D," the gradient of a function yields the direction of a vector pointing to the maximum rate of change in the slope of the function; the magnitude if the gradient is the slope of the function in that vector direction.

Similarly, in three dimensions, or "3-D," the gradient is the direction of a vector normal to the surface of the function at a given point, and the magnitude of this is the maximum rate of change of the surface function along the normal vector direction. The "Directional Derivative," or $\hat{r} \cdot \nabla f$ is the slope of the function in the direction of the <u>unit vector</u> $\hat{r}$.

The Curl

The Curl indicates a rotational component of the vector field $\bar{F}$. The specific physical meaning is tied to the physics to which it is applied, most often in the application of Maxwell's equations in electrical engineering and magnetism applications. The Curl is defined as $\nabla \times \bar{F}$ and is defined for a vector field

$$\bar{F} = F_1\hat{i} + F_2\hat{j} + F_3\hat{k} = \langle F_1, F_2, F_3 \rangle$$

and $\nabla \times \bar{F} =$

$$\begin{vmatrix} \hat{i} & \hat{j} & k \\ \dfrac{\partial}{\partial x} & \dfrac{\partial}{\partial y} & \dfrac{\partial}{\partial z} \\ F_1 & F_2 & F_3 \end{vmatrix} = \hat{i}\left(\frac{\partial F_3}{\partial y} - \frac{\partial F_2}{\partial z}\right) - \hat{j}\left(\frac{\partial F_3}{\partial x} - \frac{\partial F_1}{\partial z}\right) + \hat{k}\left(\frac{\partial F_2}{\partial x} - \frac{\partial F_1}{\partial y}\right)$$

If the vector field $\bar{F}$ is irrotational, then $\nabla \times \bar{F} = \bar{0}$ and $\bar{F}$ is a conservative vector field that has a potential function; a is conservative vector field $\bar{F}$ comes from the gradient of a scalar function $\nabla f = \bar{F}$ where f is a scalar potential function.

The Divergence

Divergence relates to a net flow of a vector field away from a point per unit volume. Formally, divergence is the net differential (outflow) rate of change of a vector field $\bar{F}$ through a spherical area per unit volume computed in the limit as the spherical radius (along with area, volume) collapses to an infinitesimally vanishing point:

$$\lim_{r \to 0} \frac{\iint_{S_r} \vec{F} \cdot d\bar{s}}{V_r} = \nabla \cdot \vec{F}$$

for surface area S_r and volume V_r.

For a vector function

$$\vec{F} = F_1 \hat{i} + F_2 \hat{j} + F_3 \hat{k} = \langle F_1, F_2, F_3 \rangle$$

The Divergence is indicated by

$$\nabla \cdot \vec{F} = \frac{\partial F_1}{\partial x} + \frac{\partial F_2}{\partial y} + \frac{\partial F_3}{\partial z}$$

Vector Field in Fluid Mechanics

The Navier–Stokes equations describe the conservation of mass, momentum and energy with regard to fluid flow. The conservation of mass is defined using divergence, where

$$\frac{\partial \rho}{dt} + \nabla \cdot (\rho \vec{v}) = 0$$

with

$\rho = \text{density } \dfrac{kg}{m^3}$

$\vec{v} = \text{velocity of a parcel of fluid m/s}$

$$\text{and} \therefore \frac{\partial \rho}{dt} \text{ has units of } \frac{kg}{m^3 s}.$$

Close inspection of this equation states that it is a balance equation between the time rate of change in the fluid density and the divergence of mass flux. If divergence of mass flux is positive, then rate of change in density is negative.

Now, "mass flux" is the amount of mass (out) passing a unit area per unit time:

$$\rho\bar{v} = \frac{kg}{m^3}\frac{m}{s} \rightarrow \frac{kg}{m^2 s}$$

The divergence of mass flux is $\nabla \cdot (\rho\bar{v})$. Given that velocity is typically expressed as a vector with individual components

$$\bar{v} = u\hat{i} + v\hat{j} + w\hat{k} \quad \text{in m/s} \quad \text{and} \quad \nabla = \frac{\partial}{\partial x}\hat{i} + \frac{\partial}{\partial y}\hat{j} + \frac{\partial}{\partial z}\hat{k}$$

Then the divergence of mass flux, $\nabla \cdot (\rho\bar{v})$ is

$$\nabla \cdot (\rho\bar{v}) = \rho\frac{\partial u}{\partial x} + \rho\frac{\partial v}{\partial y} + \rho\frac{\partial w}{\partial z}\Bigg\} \quad \text{with units of} \quad \frac{kg}{m^3}\frac{m}{s \cdot m} \rightarrow \frac{kg}{m^3 s}$$

Again, note how all units in any differential equation must match!

Gauss's Divergence Theorem

Given the definition of the divergence as net outflow per unit volume of a vector field, it can readily be seen that *Gauss' Divergence theorem* holds true: for a given vector field, the surface integral of a vector field can be converted to a volume integral of the divergence of a vector field

$$\int_S \bar{F} \cdot d\bar{s} = \int_V \nabla \cdot \bar{F}dV$$

Example

Consider the mass flux of a fluid given by

$$\bar{F}(x, y, z) = xy\hat{i} + 3yz\hat{j} + 2e^z\hat{k} = \rho\bar{v} \quad \frac{kg}{m^2 s}$$

If this fluid is contained in a Cartesian box bounded between $(0,0,0)$ and $(1,1,1)$ demonstrate the validity of Gauss's divergence theorem.

To compute total outflow rate using the computation of the net mass flux of fluid across each surface requires the application of six individual surface integrals, as follows:

$$\int_S \vec{F} \cdot d\vec{s} = \int_0^1 \int_0^1 \vec{F}(0,y,z) dy dz \cdot (-\hat{i}) + \int_0^1 \int_0^1 \vec{F}(1,y,z) dy dz \cdot (\hat{i})$$

$$+ \int_0^1 \int_0^1 \vec{F}(x,0,z) dx dz \cdot (-\hat{j}) + \int_0^1 \int_0^1 \vec{F}(x,1,z) dx dz \cdot (\hat{j})$$

$$+ \int_0^1 \int_0^1 \vec{F}(x,y,0) dx dy \cdot (-\hat{k}) + \int_0^1 \int_0^1 \vec{F}(x,y,1) dx dy \cdot (\hat{k}) = 2e^1 \frac{kg}{s}$$

Using Gauss's theorem, rather than compute each of the six surface integrals, one can simply take the divergence of the vector mass flux of fluid and integrate over the volume of the box:

$$\int_V \nabla \cdot \vec{F} \, dV$$

$$\nabla \cdot \vec{F} = \left(\frac{\partial}{\partial x} \hat{i} + \frac{\partial}{\partial y} \hat{j} + \frac{\partial}{\partial z} \hat{k} \right) \cdot \left(xy\hat{i} + 3yz\hat{j} + 2e^z \hat{k} \right)$$

$$= y + 3z + 2e^z$$

$$\Rightarrow \int_0^1 \int_0^1 \int_0^1 \left(y + 3z + 2e^z \right) dx \, dy \, dz = 2e^1 \frac{kg}{s}$$

Gauss' theorem is most often applied to convert from a surface integral to a volume integral in differential equations.

Operations in General Coordinates

The derivation of the Del operator and the various operations (the gradient, divergence, and curl) in general coordinate systems can be derived by translating scale lengths and applying the fundamental theorem of calculus. First, consider, $d\vec{r}$ a differential position vector in 3-space. This differential position vector can be defined using

$$d\vec{r} = (h_1 dq_1)\hat{q}_1 + (h_2 dq_2)\hat{q}_2 + (h_3 dq_3)\hat{q}_3$$

and "scale factors" and coordinate axis dimensions, respectively, are

$$(h_1, \quad h_2, \quad h_3) \qquad\qquad (q_1, \quad q_2, \quad q_3)$$

and the set of unit vectors along each coordinate axis dimension is defined as

$$(\hat{q}_1, \quad \hat{q}_2, \quad \hat{q}_3)$$

Also, consider a vector valued function defined in this coordinate system

$$\vec{F} = F_1\hat{q}_1 + F_2\hat{q}_2 + F_3\hat{q}_3$$

By the Fundamental Theorem of Calculus for Line Integrals, we can write

$$\oint \vec{F} \cdot d\vec{r} = \oint \frac{\partial f}{\partial q_1} h_1 dq_1 + \frac{\partial f}{\partial q_2} h_2 dq_2 + \frac{\partial f}{\partial q_3} h_3 dq_3 = f(B) - f(A) = 0$$

Where $\vec{F} = \nabla f$ from some scalar potential function $f(x)$. This means that

$$df = \nabla f \cdot d\vec{r} = \vec{F} \cdot d\vec{r} = (F_1 h_1 dq_1 + F_2 h_2 dq_2 + F_3 h_3 dq_3)$$

Then, from the *chain rule* the total differential of the potential function $f(x)$

$$df = \left(\frac{\partial f}{\partial q_1} dq_1 + \frac{\partial f}{\partial q_2} dq_2 + \frac{\partial f}{\partial q_3} dq_3 \right)$$

Since the total differential must be fully consistent, where $df = df$, then

$$\left(F_1 h_1 dq_1 + F_2 h_2 dq_2 + F_3 h_3 dq_3\right) = \left(\frac{\partial f}{\partial q_1} dq_1 + \frac{\partial f}{\partial q_2} dq_2 + \frac{\partial f}{\partial q_3} dq_3\right)$$

We can then solve for each component of the vector function $\bar{F}$ in terms of the scale factors

$$F_1 = \frac{1}{h_1}\frac{\partial f}{\partial q_1} \quad F_2 = \frac{1}{h_2}\frac{\partial f}{\partial q_2} \quad F_3 = \frac{1}{h_3}\frac{\partial f}{\partial q_3}$$

This then means that the gradient is determined from

$$\nabla f = \frac{1}{h_1}\frac{\partial f}{\partial q_1}\hat{q}_1 + \frac{1}{h_2}\frac{\partial f}{\partial q_2}\hat{q}_2 + \frac{1}{h_3}\frac{\partial f}{\partial q_3}\hat{q}_3$$

and we can write the Del operator, for general coordinate systems:

$$\nabla \equiv \frac{1}{h_1}\frac{\partial}{\partial q_1}\hat{q}_1 + \frac{1}{h_2}\frac{\partial}{\partial q_2}\hat{q}_2 + \frac{1}{h_3}\frac{\partial}{\partial q_3}\hat{q}_3$$

In Cartesian coordinates,

$$d\bar{r} = \left(h_1 dq_1\right)\hat{q}_1 + \left(h_2 dq_2\right)\hat{q}_2 + \left(h_3 dq_3\right)\hat{q}_3 = dx\,\hat{i} + dy\,\hat{j} + dz\,\hat{k}$$

and it is then clear that

$$h_1 = 1, \quad h_2 = 1, \quad h_3 = 1$$

$$q_1 = x, \quad q_2 = y, \quad q_3 = z$$

$$\hat{q}_1 = \hat{i}, \quad \hat{q}_2 = \hat{j}, \quad \hat{q}_3 = \hat{k}$$

and the gradient of a scalar function $f(x)$ is therefore

$$\nabla f = \frac{1}{1}\frac{\partial f}{\partial x}\hat{i} + \frac{1}{1}\frac{\partial f}{\partial y}\hat{j} + \frac{1}{1}\frac{\partial f}{\partial z}\hat{k} = \frac{\partial f}{\partial x}\hat{i} + \frac{\partial f}{\partial y}\hat{j} + \frac{\partial f}{\partial z}\hat{k}$$

Based on the definition of divergence, the net outflow of a vector function

$$\bar{F} = F_1\hat{i} + F_2\hat{j} + F_3\hat{k} = \langle F_1, F_2, F_3 \rangle$$

across a differential surface of dimensions $h_1\,dq_1$ $h_2\,dq_2$ $h_3\,dq_3$ per unit differential volume, where $dV = h_1\,dq_1$ $h_2\,dq_2$ $h_3\,dq_3$ ... making use of a truncated Taylor's series to determine the net flow out of each differential surface, is:

$$\begin{bmatrix} \left(F_1\,h_2 dq_2\,h_3 dq_3 + \dfrac{\partial}{\partial q_1}(F_1\,h_2 dq_2\,h_3 dq_3)dq_1 \right) - \left(F_1\,h_2 dq_2\,h_3 dq_3 \right) + \\[2ex] \left(F_2\,h_1 dq_1\,h_3 dq_3 + \dfrac{\partial}{\partial q_2}(F_2\,h_1 dq_1\,h_3 dq_3)dq_2 \right) - \left(F_2\,h_1 dq_1\,h_3 dq_3 \right) + \\[2ex] \left(F_3\,h_1 dq_1\,h_2 dq_2 + \dfrac{\partial}{\partial q_3}(F_3\,h_1 dq_1\,h_2 dq_2)dq_3 \right) - \left(F_3\,h_1 dq_1\,h_2 dq_2 \right) \end{bmatrix} / [(h_1 dq_1\,h_2 dq_2\,h_3 dq_3)]$$

This simplifies to

$$\nabla \cdot \vec{F} = \frac{1}{h_1 h_2 h_3}\left[\frac{\partial}{\partial q_1}(F_1 h_2 h_3) + \frac{\partial}{\partial q_2}(F_2 h_3 h_1) + \frac{\partial}{\partial q_3}(F_3 h_1 h_2) \right]$$

and in Cartesian coordinates yields

$$\nabla \cdot \vec{F} = \frac{\partial F_1}{\partial x} + \frac{\partial F_2}{\partial y} + \frac{\partial F_3}{\partial z}$$

Then, if $\vec{F} = \nabla f$, the divergence of the gradient of the scalar function $f(x)$ is the Laplacian, given by

$$\nabla \cdot \nabla f = \nabla^2 f = \frac{1}{h_1 h_2 h_3}\left[\frac{\partial}{\partial q_1}\left(\frac{h_2 h_3}{h_1}\frac{\partial f}{\partial q_1} \right) + \frac{\partial}{\partial q_2}\left(\frac{h_3 h_1}{h_2}\frac{\partial f}{\partial q_2} \right) + \frac{\partial}{\partial q_3}\left(\frac{h_1 h_2}{h_3}\frac{\partial f}{\partial q_3} \right) \right]$$

Operations in Spherical Coordinates

It is often convenient to choose coordinate system vectors that are orthogonal to each other; in the most general sense, if the vectors are not precisely orthogonal (but are still linearly independent), they are called *oblique* vectors. This leads to a study of Graph Theory, and is beyond the scope of this discussion. Fundamentally, we have employed a standard Cartesian orthonormal basis: as discussed, this means the vectors are mutually orthogonal vectors of unit length.

Cylindrical or spherical coordinate systems are also orthonormal, and can be directly derived from Cartesian coordinates using equivalent formulations for x, y, and z; Cylindrical and spherical coordinates are formulated to take advantage of spherical symmetry.

In this section we will focus on the spherical coordinate system. To derive operations in spherical coordinates, we begin with the position vector

$$\vec{r} = x\hat{i} + y\hat{j} + z\hat{k}$$

Where we translate this into spherical coordinates using the relationship between x, y, and z and ρ, θ, ϕ :

$$\rho \in [0, \infty), \quad \theta \in [0, \pi], \quad \phi \in [0, 2\pi]$$
$$x = \rho \sin\theta \cos\phi$$
$$y = \rho \sin\theta \sin\phi$$
$$z = \rho \cos\theta$$

And the position vector is

$$\vec{r} = \rho \sin\theta \cos\phi \, \hat{i} + \rho \sin\theta \sin\phi \, \hat{j} + \rho \cos\theta \, \hat{k}$$

Consider that we need to determine the appropriate scale factors for the spherical coordinate system; recall that from the general coordinate system, this leads to the relationship that for the spherical coordinate system, we have

$$d\vec{r} = (h_\rho d\rho)\hat{\rho} + (h_\theta d\theta)\hat{\theta} + (h_\phi d\phi)\hat{\phi}$$

Alternatively, in terms of the *total differential* of $\vec{r}$

$$d\vec{r} = \frac{\partial \vec{r}}{\partial \rho} d\rho + \frac{\partial \vec{r}}{\partial \theta} d\theta + \frac{\partial \vec{r}}{d\phi} d\phi$$

Then it is clear that

$$h_\rho \hat{\rho} = \frac{\partial \vec{r}}{\partial \rho} \qquad h_\theta \hat{\theta} = \frac{\partial \vec{r}}{\partial \theta} \qquad h_\phi \hat{\phi} = \frac{\partial \vec{r}}{d\phi}$$

Using $\vec{r} = \rho \sin\theta \cos\phi \, \hat{i} + \rho \sin\theta \sin\phi \, \hat{j} + \rho \cos\theta \, \hat{k}$ we determine that

$$\frac{\partial \vec{r}}{\partial \rho} = \sin\theta \cos\phi \, \hat{i} + \sin\theta \sin\phi \, \hat{j} + \cos\theta \, \hat{k} = h_\rho \hat{\rho}$$

$$\frac{\partial \bar{r}}{\partial \theta} = \rho \cos \theta \cos \phi \,\hat{i} + \rho \cos \theta \sin \phi \,\hat{j} + -\rho \sin \theta \,\hat{k} = h_\theta \hat{\theta}$$

$$\frac{\partial \bar{r}}{d \phi} = -\rho \sin \theta \sin \phi \,\hat{i} + \rho \sin \theta \cos \phi \,\hat{j} + 0 \,\hat{k} = h_\phi \hat{\phi}$$

Then, we can determine each scale factor by finding the magnitude of both sides:

$$h_\rho = \left| \frac{\partial \bar{r}}{\partial \rho} \right| = \sqrt{\sin^2 \theta \cos^2 \phi + \sin^2 \theta \sin^2 \phi + \cos^2 \theta}$$

Or

$$h_\rho = \sqrt{\sin^2 \theta \left(\cos^2 \phi + \sin^2 \phi \right) + \cos^2 \theta} = 1$$

In a similar manner, it can be shown that

$$h_\theta = \rho \qquad \text{and} \qquad h_\phi = \rho \sin \theta$$

For spherical coordinates, with

$$h_\rho = 1; \quad h_\theta = \rho; \quad h_\phi = \rho \sin \theta$$

The differential position vector is

$$d\bar{r} = 1 \hat{\rho} \, d\rho + \rho \hat{\theta} \, d\theta + \rho \sin \theta \hat{\phi} \, d\phi$$

Recalling that

$$\nabla = \frac{1}{h_1} \frac{\partial}{\partial q_1} \hat{q}_1 + \frac{1}{h_2} \frac{\partial}{\partial q_2} \hat{q}_2 + \frac{1}{h_3} \frac{\partial}{\partial q_3} \hat{q}_3$$

We can determine that the gradient operator for spherical coordinates is:

$$\nabla = \frac{\partial}{\partial \rho} \hat{\rho} + \frac{1}{\rho} \frac{\partial}{\partial \theta} \hat{\theta} + \frac{1}{\rho \sin \theta} \frac{\partial}{\partial \phi} \hat{\phi}$$

The Laplacian is

$$\nabla \cdot \nabla f \equiv \nabla^2 f = \frac{1}{\rho^2 \sin \theta} \left[\frac{\partial}{\partial \rho} \left(\frac{\rho^2 \sin \theta}{1} \frac{\partial f}{\partial \rho} \right) + \frac{\partial}{\partial \theta} \left(\frac{\rho \sin \theta}{\rho} \frac{\partial f}{\partial \theta} \right) + \frac{\partial}{\partial \phi} \left(\frac{\rho}{\rho \sin \theta} \frac{\partial f}{\partial \phi} \right) \right]$$

Expanding with further simplification yields

$$\nabla^2 f = \frac{1}{\rho^2 \sin\theta}\left[\sin\theta \frac{\partial}{\partial \rho}\left(\rho^2 \frac{\partial f}{\partial \rho}\right)\right] + \frac{1}{\rho^2 \sin\theta}\left[\frac{\rho}{\rho}\frac{\partial}{\partial \theta}\left(\sin\theta \frac{\partial f}{\partial \theta}\right)\right] + \frac{1}{\rho^2 \sin\theta}\left[\frac{\partial}{\partial \phi}\left(\frac{1}{\sin\theta}\frac{\partial f}{\partial \phi}\right)\right]$$

$$\nabla^2 f = \frac{1}{\rho^2}\frac{\partial}{\partial \rho}\left(\rho^2 \frac{\partial f}{\partial \rho}\right) + \frac{1}{\rho^2 \sin\theta}\frac{\partial}{\partial \theta}\left(\sin\theta \frac{\partial f}{\partial \theta}\right) + \frac{1}{\rho^2 \sin\theta}\frac{\partial^2 f}{\partial \phi^2}$$

$$\nabla^2 f = \frac{1}{\rho^2}\left(2\rho\frac{\partial f}{\partial \rho} + \rho^2 \frac{\partial^2 f}{\partial \rho^2}\right) + \frac{1}{\rho^2 \sin\theta}\left(\cos\theta \frac{\partial f}{\partial \theta} + \sin\theta \frac{\partial^2 f}{\partial \theta^2}\right) + \frac{1}{\rho^2 \sin^2\theta}\frac{\partial^2 f}{\partial \phi^2}$$

and finally

$$\nabla^2 f = \left[\frac{\partial^2 f}{\partial \rho^2} + \frac{2}{\rho}\frac{\partial f}{\partial \rho} + \frac{\cot\theta}{\rho^2}\frac{\partial f}{\partial \theta} + \frac{1}{\rho^2}\frac{\partial^2 f}{\partial \theta^2} + \frac{1}{\rho^2 \sin^2\theta}\frac{\partial^2 f}{\partial \phi^2}\right]$$

General Differential Volume Element and the Jacobian

Recall that the general differential volume element can be computed by combining scale factors and differentials in each coordinate system:

$$dV = h_1 dq_1 h_2 dq_2 h_3 dq_3$$

Considering the spherical coordinate system derived in the previous section

$$h_1 = 1 \qquad dq_1 = d\rho$$
$$h_2 = \rho \qquad dq_2 = d\theta$$
$$h_3 = \rho\sin\theta \qquad dq_3 = d\phi$$

Then we have $dV = 1 d\rho\,\rho\,d\theta\,\rho\sin\theta\,d\phi = \rho^2 \sin\theta\,d\rho\,d\theta\,d\phi$

Related to this is the *Jacobian*, a special type of determinant, where the general coordinate system is related to the translation from the Cartesian system:

$$(\text{Jacobian}) = \det \begin{bmatrix} \dfrac{\partial x}{\partial q_1} & \dfrac{\partial y}{\partial q_1} & \dfrac{\partial z}{\partial q_1} \\[3mm] \dfrac{\partial x}{\partial q_2} & \dfrac{\partial y}{\partial q_2} & \dfrac{\partial z}{\partial q_2} \\[3mm] \dfrac{\partial x}{\partial q_3} & \dfrac{\partial y}{\partial q_3} & \dfrac{\partial z}{\partial q_3} \end{bmatrix}$$

And the differential volume is computed from

$$dV = (\text{Jacobian}) \cdot dq_1 \, dq_2 \, dq_3$$

Applied to spherical coordinates, it can be shown that using

$$x = \rho \sin\theta \cos\phi$$
$$y = \rho \sin\theta \sin\phi \qquad \text{and} \qquad q_1 = \rho, \ q_2 = \theta, \ q_3 = \phi$$
$$z = \rho \cos\theta$$

Where the Jacobian = $\rho^2 \sin\theta$ and $dV = \rho^2 \sin\theta \, d\rho \, d\theta \, d\phi$

Surface Integrals

Surface integrals arise from the need to compute the area of a surface described by a function described as $z = f(x, y)$:

$$S = \iint_S dS$$

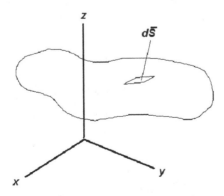

This figure depicts a surface with a differential surface element dS

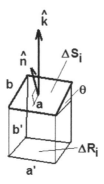

A dS geometry construct

First, we must considering a differential surface element dS. This can be derived as follows. Let

$$a = a' = \Delta x$$

$$b = \frac{b'}{\cos\theta} = \frac{\Delta y}{\cos\theta} = \frac{\Delta y}{\hat{n} \cdot \hat{k}}$$

$$\Delta S_i = \frac{\Delta R_i}{\cos\theta}$$

$$dS = \lim_{\substack{\Delta x \to 0 \\ \Delta y \to 0}} \Delta S_i \to \frac{dxdy}{\hat{n} \cdot \hat{k}}$$

Then given the surface equation cast as the function $z = f(x, y)$ is recast as $h(x,y,z) = -f(x,y) + z = 0$

Then the normal vector to $h(x,y,z)$ is $\vec{N} = \nabla h(x, y, z)$

$$\vec{N} = \nabla h(x, y, z) = \frac{\partial h}{\partial x}\hat{i} + \frac{\partial h}{\partial y}\hat{j} + \frac{\partial h}{\partial z}\hat{k}$$

So that

$$\left[\vec{N} = -\frac{\partial f}{\partial x}\hat{i} + -\frac{\partial f}{\partial y}\hat{j} + 1\hat{k} \right]$$

Then the *unit* normal vector is

$$\hat{n} = \frac{\vec{N}}{|\vec{N}|} = \frac{-\dfrac{\partial f}{\partial x}\hat{i} + -\dfrac{\partial f}{\partial y}\hat{j} + 1\hat{k}}{\sqrt{\left(\dfrac{\partial f}{\partial x}\right)^2 + \left(\dfrac{\partial f}{\partial y}\right)^2 + 1^2}}$$

and then

$$\hat{n} \cdot \hat{k} = \frac{1}{\sqrt{\left(\dfrac{\partial f}{\partial x}\right)^2 + \left(\dfrac{\partial f}{\partial y}\right)^2 + 1}}$$

So that since

$$dS = \frac{dx\,dy}{\hat{n} \cdot \hat{k}}$$

we have the formulation for the differential surface area

$$dS = \left(\sqrt{\left(\dfrac{\partial f}{\partial x}\right)^2 + \left(\dfrac{\partial f}{\partial y}\right)^2 + 1}\right) dx\,dy$$

Whereupon the total surface area is

$$S = \iint_S \sqrt{\left(\dfrac{\partial f}{\partial x}\right)^2 + \left(\dfrac{\partial f}{\partial y}\right)^2 + 1}\, dx\,dy$$

Surface integrals arise from the need to compute the net contribution from a vector field, such as a neutron current, as it crosses a non-orthogonal surface.

$$\iint_S \vec{J} \cdot \hat{n}\, dS$$

Partial Differential Equations

Partial Differential Equations typically describe engineering phenomena and processes. The nuclear engineer must deal with heat transfer, fluid flow, criticality and neutron economy, radiation shielding and dosimetry, structural integrity, etc, all of which can be described using a PDE.

A Problem involving the solution of a PDE is "well posed" if a solution exists, a solution is unique, and the solution is based upon reliable data or physical constants. So called "boundary value problems" are steady state problems that contain boundary conditions only, whereas "initial boundary value problems" contain initial and boundary conditions.

PDEs are often categorized as elliptic, parabolic, or hyperbolic types of equations.

- *Elliptic* equations are typically steady state equations, and a solution is driven by fixed boundary values. Solution is rendered by applying boundary conditions.
- *Parabolic* equations are typically driven by time varying boundary conditions, where the time varying boundary marches the solution to ultimately arrive at steady state values. Solutions are rendered by applying initial and boundary conditions.
- *Hyperbolic* equations typically yield solutions that are characteristic of waves or vibrating systems. Solutions are rendered by applying initial and boundary conditions.

The general basic formulation for a PDE is given by

$$a\frac{\partial^2 \phi}{\partial x^2} + b\frac{\partial^2 \phi}{\partial x \partial y} + c\frac{\partial^2 \phi}{\partial y^2} + d\frac{\partial \phi}{\partial x} + f\frac{\partial \phi}{\partial y} + g\phi(x, y) = h$$

Where a, b, c, d, f, g, and h can be functions of x, y, or constants.

The behavior of the PDE is defined by the coefficients a, b, and c, and can be determined by applying the *discriminant*.

The Discriminant

The discriminant is used to help classify the partial differential equation as elliptic, parabolic, or hyperbolic:

$$D = b^2 - 4ac < 0 \qquad \text{elliptic}$$
$$D = b^2 - 4ac = 0 \qquad \text{parabolic}$$
$$D = b^2 - 4ac > 0 \qquad \text{hyperbolic}$$

The value of the discriminant will determine the behavior of the PDE.

Example

Consider LaPlace's equation:

$$\nabla^2 \phi = 0$$
$$\frac{\partial^2 \phi}{\partial x^2} + \frac{\partial^2 \phi}{\partial y^2} = 0$$
$$a = 1 \qquad b = 0 \qquad c = 1$$
$$D = b^2 - 4ac = -4$$

Because the Discriminant D is less than zero, LaPlace's equation is Elliptic.

Duplicative Rule: when encountering duplicative terms, one of these can be omitted when attempting to classify a PDE.

Therefore, when encountering an equation with mixed space and time variables, classify the PDE by omitting one of the space variables if the derivatives are of the same order, then pair the space variable term with a time variable term. For example, consider the time dependent heat conduction equation:

$$\rho c \frac{\partial T}{\partial t} = k \left[\frac{\partial^2 T}{\partial x^2} + \frac{\partial^2 T}{\partial y^2} \right]$$

The space variable is 2nd order and duplicative for classification purposes. Therefore, we eliminate the second partial in y and rearrange to obtain

$$k \frac{\partial^2 T}{\partial x^2} = \rho c \frac{\partial T}{\partial t}$$

$$a = k \qquad b = 0 \qquad c = 0$$

$$b^2 - 4k(0) = 0$$

Therefore, based on the value of the Discriminant D the equation is *Parabolic*.

Other problems:

$$3 \frac{\partial^2 u}{\partial x^2} + 2 \frac{\partial^2 u}{\partial x \partial y} + 5 \frac{\partial^2 u}{\partial y^2} + x \frac{\partial u}{\partial y} = 0$$

$$\frac{\partial^2 u}{\partial x^2} + y \frac{\partial^2 u}{\partial y^2} = 0$$

$$\alpha^2 u_{xx} - u_{tt} = 0$$

$$u_{xx} + u_{yy} - u_t = 0$$

Solution i): $4 - 4(3)(5) = -56$,   The equation is Elliptic.

Solution ii): $b = 0, a = 1, c = y$

$$b^2 - 4ac = 0 - 4ay = ?$$

$$\Rightarrow -4ay = 0$$

This equation is Elliptic for $y > 0$, Parabolic for $y = 0$, Hyperbolic for $y < 0$

Solution iii): $a = \alpha^2 \quad b = 0 \quad c = -1$

$$0^2 - 4\alpha^2(-1) = 4\alpha^2 > 0 \,,$$

This equation is Hyperbolic

Solution iv): Applying the Duplicative Rule, eliminate second partial with respect to y from consideration

$$a = 1 \quad b = 0 \quad c = 0 \qquad 0^2 - 4(1)(0) = 0$$

... This equation is Parabolic

Boundary Conditions

Dirichlet Boundary Conditions have a fixed value:

$$u(x, y)\big|_{(x_0, y_0)} = c_0$$

Neumann Boundary Conditions specify a derivative:

$$\frac{\partial u}{\partial x}\bigg|_{(x_0, y_0)} = c_0$$

Mixed or *Robin's Conditions* are a combination of Dirichlet and Neumann conditions:

$$\alpha u(x, y) + \beta \frac{\partial u}{\partial x}\bigg|_{(x_0, y_0)} = c_0$$

Initial conditions specify an initial "state point" at some initial time:

$$u(x, y, t)\big|_{t \Rightarrow t_0} = d_0$$

PDE Solutions and the Sturm–Liouville Problem

Partial Differential Equations (PDE's) can be solved in a variety of ways, including by Separation of variables without or with a change in variables, eigenfunction expansion, numerical methods, the application of integral transforms, variational methods, and others. In this text, our analytical solution efforts will focus on the popular solution technique called *separation of variables*. We will also explore the method of eigenfunction expansion, and numerical methods.

One requirement of Separation of Variables is that homogeneous boundary conditions can be made available to enable one to render solutions. The following are important facts on separation of variables.

- *Separation of variables* assumes that the variables are in fact *separable* relative to functions of the respective independent variables in the problem. For example, a function $\psi(x, y, z) = u(x)v(y)w(z)$ is assumed to be separable into separate functions u, v, and w.

- Separation of variables relies on the availability of linear homogeneous boundary conditions and the superposition principle. The principle of superposition states that multiple solutions of a linear PDE with various isolated boundary conditions applied with homogeneous boundary conditions can be summed to form a complete solution to the PDE. This is likely best illustrated with an example.

- When we separate variables, we obtain a set of ordinary differential equations (ODE's). The ODEs that are separated out from the PDEs are known as *Liouville equations*. Determining the solutions and properties of solutions, specifically the eigenvalues and eigenvectors the satisfy the differential equation, is known as the *Sturm–Liouville problem*.

- Often solutions yield infinite series solutions using orthogonal functions. Because these solutions using infinite series of orthogonal functions form *complete solutions* to the differential equations they converge to a unique solution given specific boundary/initial conditions.

- The solutions are formed in a function space called a *Hilbert Space*—this is an infinitely dimensional, complete, inner product function space. Expansions in Hilbert Space (in terms of infinite-dimensional sets of orthogonal functions) are a common way of representing many types of differential equations. *Trigonometric Fourier Series* are the simplest example of expansions in Hilbert space used to solve differential equations.

Chapter 11

Applied Solution Methods—Part 2: PDEs and Heat Transfer

In this chapter, we consider both ODEs and partial differential equations, or PDEs, solved using common methods encountered in nuclear engineering applications and related problems of interest. Fundamentally, PDEs are typically solved by applying methods similar to those used for ordinary differential equations, but with the added complexity of separating multivariate conditions, and applying orthogonality to yield a linearly independent solution basis. For this reason, we focus on solving the heat equation. Then, after covering essential concepts here, we will consider problems involving neutron transport and diffusion theory.

The Heat Equation

Heat transfer is a topic that is second nature to nuclear engineers; in particular. any practical fission reactor system requires thoughtful engineering design when it comes to heat removal; in most cases, the heat energy derived from fission systems is the most important product of a reactor. Here we again apply Fourier's law of conduction, which describes the flow of energy by conduction, where molecules vibrate together to transfer energy:

$$\vec{q}'' = -k\nabla T = -k\left(\frac{\partial T}{\partial x}\hat{i} + \frac{\partial T}{\partial y}\hat{j} + \frac{\partial T}{\partial z}\hat{k}\right)$$

where $\vec{q}''$ is the vector of energy per unit time per unit area (W/m^2), k is the solid conductivity (W/(m K)), driven by a negative temperature gradient (K/m). Since the gradient term makes this a vector, we then dot this with an area vector with a unit surface normal vector to track the flow of energy across a surface area.

To derive the heat conduction equation using a balance of energy over a differential volume element in Cartesian coordinates. The differential volume in this case is anchored with an origin at zero, spanning differential lengths in all three dimensions:

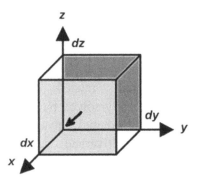

Differential volume element

For the moment, we will consider insulated conditions along y and z; that is, heat can only flow along the x-axis. To construct the heat equation, we must balance all of the energy flow in the differential "voxel."

Therefore, we begin with the rate that energy enters the voxel along the rear side at $x = 0$, noting that the differential surface area normal vector for energy entering the voxel is along the positive direction:

$$-k\,dydz\,\hat{i} \cdot \frac{\partial T}{\partial x}\bigg|_0 \hat{i}$$

The rate that heat energy is generated inside the voxel is based on the product of the volumetric heat generation rate q''' (W/m^3) and the differential volume

$$q'''\,dx\,dy\,dz$$

The rate energy leaves out of the front surface along x at a distance of *dx* is based on the value at $x = 0$ expanded in a truncated Taylor's series from the origin:

$$-k\,dydz\,\hat{i} \cdot \frac{\partial}{\partial x}\left(T\big|_0 + \frac{\partial T}{\partial x}\bigg|_0\,dx\right)\hat{i}$$

Apart from the flow of energy, the simple net rate at which energy is stored is based simply on the density ρ (kg/m^3), heat capacity c_p (J/(kg K)), and time rate of change of the temperature of the solid material composing the voxel, and is given by:

$$\rho c_p \frac{\partial T}{\partial t}\,dxdydz$$

Now, adding the energy entering the voxel, adding the energy generated within the voxel, and subtracting the energy leaving the voxel, at any instant in time, must equal the net rate at which energy is stored in the voxel. Therefore, we obtain:

$$-k dydz\frac{\partial T}{\partial x} + q'''dxdydz - -kdydz\left(\frac{\partial T}{\partial x} + dx\frac{\partial^2 T}{\partial x^2}\right) = \rho c_p \frac{\partial T}{\partial t}dxdydz$$

Simplifying, we obtain

$$+ k\,dxdydz\frac{\partial^2 T}{\partial x^2} + q'''dxdydz = \rho c_p \frac{\partial T}{\partial t}dxdydz$$

Cancelling the differential volume on both sides, we obtain

$$k\frac{\partial^2 T}{\partial x^2} + q''' = \rho c_p \frac{\partial T}{\partial t}$$

A special physical constant is called the Thermal Diffusivity, and it is defined as

$$\alpha = \frac{k}{\rho c_p}$$

The Thermal Diffusivity is the ratio of thermal conductivity to thermal capacity; thermal energy diffuses rapidly through substances with a high thermal diffusivity. Comparison of typical values of thermal diffusivity at room temperature for a few materials is:

Values of Thermal Diffusivity for Various Metals

Material	α sq ft/h
Al	3.33
Cu	4.42
Au	4.68
Fe	0.7
Ni	0.92

For this reason, it is easy to understand why heat sinks are made from copper or aluminum, and not iron, steel, or nickel. Gold would be ideal, except for the expense and weight in many applications. Using the thermal diffusivity, the heat equation is then

$$\frac{\partial^2 T}{\partial x^2} + \frac{q'''}{k} = \frac{1}{\alpha}\frac{\partial T}{\partial t}$$

If we were to track heat entering and exiting along all three axles and all six surfaces of the voxel, we would obtain

$$k\left(\frac{\partial^2 T}{\partial x^2} + \frac{\partial^2 T}{\partial y^2} + \frac{\partial^2 T}{\partial z^2}\right) + q''' = \rho c_p \frac{\partial T}{\partial t}$$

Which can be re-written as

$$k\nabla^2 T + q''' = \rho c_p \frac{\partial T}{\partial t}$$

And in using the thermal diffusivity, the heat conduction equation is

$$\nabla^2 T + \frac{q'''}{k} = \frac{1}{\alpha}\frac{\partial T}{\partial t}$$

NUCLEAR APPLICATION: 1-D Radial Heat Conduction

For very long, thin rods, such as nuclear fuel rods generating heat from nuclear fission, most of the heat transfer occurs radially. Consider a fuel rod radial cross section for a nuclear fuel rod that has an annular center, with nuclear fuel bounded by an inner radius r_i and r_o with an outer radius of fuel *cladding* (used to encapsulate the fuel) at position r_c.

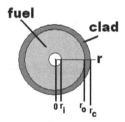

Cross section cutaway of a nuclear fuel rod

Assuming a constant power level, and assuming heat only flows radially, in cylindrical coordinates the steady state heat conduction equation is

$$\frac{1}{r}\frac{d}{dr}\left(r\frac{dT}{dr}\right)+\frac{q_f{}'''}{k_f}=0$$

Where k_f is the fuel conductivity. Assuming there is a critical reactor that is generating power, the volumetric heat generation term in this application can be determined by the local fission density. The fission density comes from the product of the local neutron flux, the fission cross section, and energy per fission:

$$q_f{}'''\left(\frac{\text{W}}{\text{cm}^3}\right)=\phi\left(\frac{\text{n}}{\text{cm}^2\text{s}}\right)\sigma_f\left(\frac{\text{fissions}\cdot\text{cm}^2}{\text{cm}^3}\right)\overline{E}_f\left(\frac{\text{J}}{\text{fission}}\right)$$

Here we assume the fission density is constant. We can solve the heat equation by multiplying both sides by r and integrating, where

$$r\frac{dT}{dr}=\frac{-q_f{}'''r^2}{2k_f}+C_1$$

Multiplying both sides by r:

$$\frac{dT}{dr} = \frac{-q_f'''r}{2k_f} + \frac{C_1}{r}$$

Applying the boundary condition that there is no heat transfer (zero temperature gradient) at the inner radius, and according to Fourier's law this also means that the temperature gradient is zero (note: this also implies, by the fundamental theorem of calculus, the this is a local maximum in temperature, which makes physical sense):

$$\frac{dT}{dr}\Big|_{r \to r_i} = 0$$

This enables us to resolve the constant of integration:

$$C_1 = \frac{q_f'''r_i^2}{2k_f}$$

Integrating both sides of the equation again

$$\int \frac{dT}{dr}\, dr = \int \frac{q_f'''}{2k_f}\left(r_i^2\frac{1}{r} - r\right)dr$$

We then get

$$T(r) = \frac{q_f'''}{2k_f}\left(r_i^2\ln(r) - \frac{r^2}{2}\right) + C_2$$

Applying a boundary condition so that

$$T(r_i) = T_i$$

allows us to solve for the constant an obtain an expression for the temperature across the fuel region. The Temperature at the outer boundary of the fuel is

$$T(r_o) = T_o$$

The temperature difference between the inner and outer fuel radius is then

$$T_i - T_o = \frac{q_f'''}{2k_f}\left(\frac{1}{2}(r_o^2 - r_i^2) - r_i^2\ln(\frac{r_o}{r_i})\right)$$

To solve for the temperature difference from the outer edge of the fuel to the outer edge of the cladding, we can directly integrate Fourier's law of conduction. Heat is conducted across the fuel surface along a unit vector $\hat{r}$ according to Fourier's law:

$$q_f'' = -k_f \frac{dT}{dr}\big|_{r \to r_o}$$

where q_f'' is the heat flux out of the fuel surface in W/cm^2. Heat exiting the fuel surface is then conducted through the clad along $\hat{r}$ according to:

$$q_c'' = -k_c \frac{dT}{dr}\big|_{r \to r_c}$$

where q_c'' is the heat flux leaving out of the cladding in W/cm^2, and k_c is the cladding conductivity. Then the heat transferred *must* be conserved according to the following energy balance relationships between the volumetric, areal (heat flux), and linear heat rates (where L represents a length of fuel):

$$q_f''' (\pi(r_o^2 - r_i^2)L) = q_f'' (2\pi r_o L) = q_c'' (2\pi r_c L) = q_c' L$$

Fourier's law of conduction may be applied anywhere heat is transferred across a boundary; within the region occupied by the cladding, the power expressed as a linear heat rate (W/m) from the nuclear fuel moves across the clad region according to

$$q_c' = -k_c (2\pi r) \frac{dT}{dr}$$

Integrating this equation on both sides as a separable equation with definite integral forms, we have

$$\int_{r_o}^{r_i} \frac{dr}{r} = \int_{T_o}^{T_c} \frac{(-2\pi k_c)}{q_c'} dT$$

Which results in the expression for temperature difference between the outer surface of the fuel and the clad:

$$T_o - T_c = \frac{q_c{'}\ln(\frac{r_c}{r_o})}{2\pi k_c}$$

Solution to the 2-D Heat Conduction Equation: Separation of Variables

Here we apply a separation of variables solution method to the heat conduction equation in (x,y) geometry. Consider a flat plate with insulated surfaces along z, so that there is no conductive heat flow along , and all conditions are steady state.

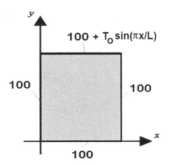

In this case, heat conduction equation becomes a boundary value problem. The heat conduction equation is

$$\nabla^2 T + \frac{q'''}{k} = \frac{1}{\alpha}\frac{\partial T}{\partial t}$$

With a specification of steady state, $\therefore \dfrac{\partial T}{\partial t} \to 0$

Without internal heat generation, $\therefore q''' \to 0$, and we then have $\nabla^2 T = 0$; with 2-D Cartesian geometry, we have

$$\nabla^2 \equiv \frac{\partial^2}{\partial x^2} + \frac{\partial^2}{\partial y^2}.$$

So that $T(x,y)$ must be solved for with the PDE to be solved as

$$\frac{\partial^2 T}{\partial x^2} + \frac{\partial^2 T}{\partial y^2} = 0$$

Next, we apply separation of variables (SOV). In doing so, we first assume that

$$T(x, y) = u(x)v(y)$$

Then, we substitute separated variables into the PDE. Next, we divide the entire equation by $T(x,y)$ to obtain

$$\left[\frac{1}{uv} (u''v + v''u = 0) \right]$$

And then

$$\frac{u''}{u} + \frac{v''}{v} = 0$$

since there are (x,y) independent dimensions, they can only ever be equal and opposite if both are equal to constants which cancel in this case. Therefore, with the use of a separation constant μ

$$-\frac{u''}{u} = \mu = +\frac{v''}{v} \leftarrow$$

Now, we must consider the following possibilities… $(\lambda > 0)$ in each case
1) $\quad \mu = -\lambda^2$
2) $\quad \mu = 0$
3) $\quad \mu = +\lambda^2$

First, consider $\mu = -\lambda^2$. Therefore,

$$-\frac{u''}{u} = -\lambda^2 \qquad \frac{v''}{v} = -\lambda^2$$

$$u'' - \lambda^2 u = 0 \qquad\qquad\qquad\qquad v'' + \lambda^2 v = 0$$

characteristic polynomial, using $u = e^{mx}$ and $v = e^{ny}$ are:

$$m^2 - \lambda^2 = 0 \qquad\qquad n^2 + \lambda^2 = 0$$

$$\therefore m = \pm \lambda \qquad\qquad n = \pm i\lambda$$

$$u(x) = \tilde{c}_1 e^{\lambda x} + \tilde{c}_2 e^{-\lambda x} \qquad v(y) = c_3 \sin(\lambda y) + c_4 \cos(\lambda y)$$

Or

$$u(x) = c_1 \sinh(\lambda x) + c_2 \cosh(\lambda x)$$

Ordinarily, this would be acceptable; since we could couple these linearly independent solutions; however, since a boundary condition has a sine function in x:

$$T(x, w) = 100 + T_0 \sin\left(\frac{\pi x}{L}\right)$$

we cannot choose this arrangement—we need to consider alternative solutions. Therefore, we eliminate $\mu = -\lambda^2$ as a possibility for the separation constant. Next, consider $\mu = 0$. We get:

$$-\frac{u''}{u} = 0 \qquad\qquad \frac{v''}{v} = 0$$

$$u'' = 0 \qquad\qquad v'' = 0$$

$$u' = c_1 \qquad\qquad v' = c_3$$

$$u(x) = c_1 x + c_2 \qquad v = c_3 y + c_4$$

We obtain

$$T(x, y) = (c_1 x + c_2)(c_3 y + c_4)$$

This could not represent

$$T(x, w) = 100 + T_0 \sin\left(\frac{\pi x}{L}\right)$$

Therefore, we eliminate $\mu = 0$ as a possibility for the separation constant. It is of note that sometimes, if boundary conditions are not made to be homogeneous for key boundaries (e.g. non-homogeneous boundaries), the linear equation can play a role in these cases. However, the linear solution has no meaning in this case...we will save other problems where the linear solution is the steady state portion, and a transient portion in time dependent equations (parabolic). Recall elliptic equations describe (typically) steady state boundary value problems.

Finally, consider $\mu = +\lambda^2$

$$-\frac{u''}{u} = \lambda^2 \qquad\qquad \frac{v''}{v} = \lambda^2$$

$$u'' + \lambda^2 u = 0 \qquad\qquad v'' - \lambda^2 v = 0$$

Then we obtain

$$u(x) = c_1 \cos(\lambda x) + c_2 \sin(\lambda x) \qquad\qquad v(y) = c_3 \sinh(\lambda y) + c_4 \cosh(\lambda y)$$

Upon inspection, this arrangement can uniquely support the boundary condition along the top edge of the plate.

$$T(x,y) = u(x)v(y)$$
$$T(x,y) = \left(c_1 \cos(\lambda x) + c_2 \sin(\lambda x)\right)\left(c_3 \sinh(\lambda y) + c_4 \cosh(\lambda y)\right)$$

Now, to simplify the solution procedure, we propose a shifted temperature: P(x,y). Therefore,

$$\left[P(x,y) + 100 = T(x,y)\right]$$

Boundary conditions that apply in terms of $P(x,y)$ are now

1) $P(0,y) = 0$
2) $P(x,0) = 0$
3) $P(L,y) = 0$
4) $P(x,w) = T_0 \sin\left(\frac{\pi x}{L}\right)$

Note: this gives us "homogeneous boundary conditions" by which separation of variables can yield a solution.

Applying boundary conditions

$$1) \quad P(0, y) = 0 \quad \therefore c_1 \to 0$$
$$2) \quad P(x,0) = 0 \quad \therefore c_4 \to 0$$

Therefore, by applying BCs 1) and 2)

$$P(x, y) = (c_2 \sin(\lambda x))(c_3 \sinh(\lambda y))$$
$$or \quad c \leftarrow (c_2 \cdot c_3)$$
$$P(x, y) = c \sin(\lambda x) \sinh(\lambda y)$$

Now apply BC 3)

$$P(L, y) = 0$$

$$P(L, y) = c \sin(\lambda L) \sinh(\lambda y) = 0$$

The only meaningful solution comes from $\sin(\lambda L) = 0$ is

$$\lambda \to \lambda_n = \left(\frac{n\pi}{L}\right)$$

satisfies this BC for $n = 1,2,3\ldots$ and $c \to c_n$. Then we have

$$P(x, y) = \sum_{n=1}^{\infty} c_n \sin\left(\frac{n\pi x}{L}\right) \sinh\left(\frac{n\pi y}{L}\right)$$

We are now ready to apply the final BC. Note the values of

$$\lambda_n = \frac{n\pi}{L}$$

are the *eigenvalues* and

$$\sin\left(\frac{n\pi x}{L}\right)$$

are *eigenfunctions* of the heat equation, a PDE that is a "Sturm–Liouville" problem.

Starting from

$$P(x, y) = \sum_{n=1}^{\infty} c_n \sin\left(\frac{n\pi x}{L}\right) \sinh\left(\frac{n\pi y}{L}\right)$$

We note that BC 4) states that

$$P(x, w) = T_0 \sin\left(\frac{\pi x}{L}\right)$$

So that we can write

$$\left[P(x, w) = T_0 \sin\left(\frac{\pi x}{L}\right) = \sum_{n=1}^{\infty} c_n \sin\left(\frac{n\pi x}{L}\right) \sinh\left(\frac{n\pi w}{L}\right) \right]$$

How do we determine c_n? The answer is clear: we now apply orthogonality by multiplying both sides of the equation by

$$\sin\left(\frac{m\pi x}{L}\right)$$

and integrate both sides from [0,L].

$$\int_0^L T_0 \sin\left(\frac{\pi x}{L}\right) \sin\left(\frac{m\pi x}{L}\right) dx = \int_0^L \sum_{n=1}^{\infty} c_n \sin\left(\frac{n\pi x}{L}\right) \sin\left(\frac{m\pi x}{L}\right) \sinh\left(\frac{n\pi w}{L}\right) dx$$

By orthogonality of trigonometric functions, and inspection of the left side of the equation, we note that ($m = 1$) is only term that yields a non-zero result.

Since we must have $m = 1$ from the left side, we subsequently have the only surviving term (again due to orthogonality) be $n = 1$, and all other terms are zero. Therefore, we have

$$T_0 \int_0^L \sin^2\left(\frac{\pi x}{L}\right) dx = \sinh\left(\frac{\pi w}{L}\right) \cdot c_1 \int_0^L \sin^2\left(\frac{\pi x}{L}\right) dx$$

and the constant can be therefore determined:

$$\therefore \left[c_1 = \frac{T_0}{\sinh\left(\dfrac{\pi w}{L}\right)} \right]$$

Therefore:

$$P(x, y) = \frac{T_0}{\sinh\left(\dfrac{\pi w}{L}\right)} \sin\left(\frac{\pi x}{L}\right) \sinh\left(\frac{\pi y}{L}\right)$$

Recalling that we obtain the final solution to the problem $T(x, y) = 100 + P(x, y)$

$$\left[T(x, y) = 100 + \frac{T_0}{\sinh\left(\dfrac{\pi w}{L}\right)} \sin\left(\frac{\pi x}{L}\right) \sinh\left(\frac{\pi y}{L}\right) \right]$$

A More General Boundary Condition

Now we consider the same problem, however, instead of the boundary condition used for BC4, we apply a more "generic" boundary condition that is only some function of x along the top of the plate, so that BC4 becomes simply

$$P(x, w) = h(x)$$

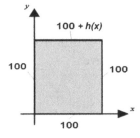

General top boundary condition

In this case, we would return to the general form of the equation as before:

$$P(x, y) = \sum_{n=1}^{\infty} c_n \sin\left(\frac{n\pi x}{L}\right) \sinh\left(\frac{n\pi y}{L}\right)$$

So that we can write

$$\left[P(x, w) = h(x) = \sum_{n=1}^{\infty} c_n \sin\left(\frac{n\pi x}{L}\right) \sinh\left(\frac{n\pi w}{L}\right) \right]$$

As before, we determine c_n by applying orthogonality, multiplying both sides of the equation by

$$\sin\left(\frac{m\pi x}{L}\right)$$

and integrate both sides from $[0, L]$:

$$\int_0^L h(x) \sin\left(\frac{m\pi x}{L}\right) dx = \int_0^L \sum_{n=1}^{\infty} c_n \sin\left(\frac{n\pi x}{L}\right) \sin\left(\frac{m\pi x}{L}\right) \sinh\left(\frac{n\pi w}{L}\right) dx$$

Left with the general definition of the boundary condition, we note that (m = n) is required to yield a non-zero result by orthogonality of trigonometric functions; therefore, the sum is no longer needed and c_n can be directly determined for this more general boundary condition to be:

$$\left[c_n = \frac{2}{L} \operatorname{csch}\left(\frac{n\pi w}{L}\right) \int_0^L h(x) \sin\left(\frac{n\pi x}{L}\right) dx \right]$$

The final solution is then complete.

$$\frac{\partial^2 T}{\partial x^2} + \frac{\partial^2 T}{\partial y^2} = 0$$

Example

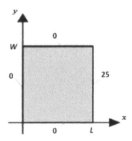

Schematic of Example problem with 25°C temperature on the right boundary

Considering the simple steady state heat transfer problem with a constant 25°C boundary condition on the right boundary. Using Separation of Variables, the solution to the heat equation is

$$T(x, y) = \sum_{n=1}^{\infty} c_n \sin\left(\frac{n\pi y}{W}\right) \sinh\left(\frac{n\pi x}{W}\right)$$

$$c_n = \frac{2}{W} \operatorname{csc} h\left(\frac{n\pi L}{W}\right) \int_0^W 25 \sin\left(\frac{n\pi y}{W}\right) dy$$

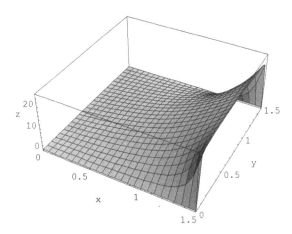

With *L=1.5 m, W=1.5m* and *n=50* terms yields the temperature profile shown.

SuperPosition Principle

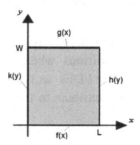

Geometry to consider for superposition

For any linear partial differential equation, the superposition principle states that if we can solve for a number of different solutions to the PDE, then the sum of the individual solutions is also a solution to the PDE. Therefore, we can use this fact to isolate the non-homogeneous boundary conditions with separation of variables to yield the solution to the differential equation. For example, consider the following problem:

$$\frac{\partial^2 T}{\partial x^2} + \frac{\partial^2 T}{\partial y^2} = 0$$

Therefore, since this is a linear PDE, we can solve this problem by obtaining a solution for each of the following cases with homogeneous conditions strategically applied and summing the solutions together:

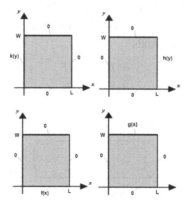

Illustration of superposition

Based on this example, there is a fundamental axiom that we must pay attention to when solving linear partial differential equations:

Axiom for Linear PDEs: One must always consider the application of **homogeneous boundary conditions** when we apply the method of separation of variables to linear PDEs, as the solutions that result from the homogeneous conditions contribute to the **complete solution** of the problem.

This axiom is clearly demonstrated in the solution of the time dependent heat transfer problem in the next section.

Solution to the Time Dependent Heat Conduction Equation

Now we will consider the solution to the time dependent heat conduction equation:

$$\nabla^2 T + \frac{q'''}{k} = \frac{1}{\alpha} \frac{\partial T}{\partial t}$$

Recall that we considered radial conduction out of a nuclear fuel rod with internal heat generation earlier in the chapter; to present a simple scenario, we will consider a rod without any internal heat generation, so $q''' = 0$. Moreover, we will consider transmission of heat along one dimension, and specifically we will consider a bar where heat is only conducted out of the two axial ends of the bar; therefore the bar is insulated along its length. The the heat equation, which is a linear partial differential equation, becomes

$$\frac{\partial^2 T}{\partial x^2} = \frac{1}{\alpha} \frac{\partial T}{\partial t}$$

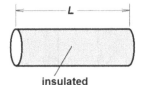

insulated

Radially insulated rod of length L

This equation is *parabolic*, which means the solution will march from an initially unstable condition to converge to a stable steady condition, driven by the boundary conditions. Next, for this case the bar of length L is initially at a fixed temperature of $2T_0$ and that at time $t = 0$, one end of the bar is suddenly held at zero degrees, while the other end of the bar is suddenly held at T_0. In equation form, these *Initial* and *Boundary conditions* are:

1) $T(x,0) = 2T_0$ (Initial condition)

2) $T(0,t) = 0$

3) $T(L,t) = T_0$

We then apply Separation of Variables with the use of a separation constant μ:

$$T(x,t) = u(x)v(t) \quad \text{and} \quad \frac{u''}{u} = \frac{1}{\alpha}\frac{v'}{v} = \mu$$

Now...we must consider the following possibilities... $(\lambda > 0)$ in each case. Since there is a spatial component and a time component that will result for each separation constant alternative, it is customary to first consider the spatial component:

Evaluate $\mu = +\lambda^2$. For the spatial component, this leads to

$$u'' - \lambda^2 u = 0$$

This directly results in hyperbolic solutions, and these are *not* consistent with either homogeneous or the given boundary conditions.

Evaluate $\mu = 0$. For the spatial component, this leads directly to

$$\frac{u''}{u} = 0 = \mu$$

$$u(x) = A_1 x + A_2$$

$$u(0) = 0 \quad \text{and} \quad u(L) = T_0 \quad \text{results in}$$

$$u_1(x) = \frac{T_0 x}{L}$$

This value of the separation constant therefore yields a *linear temperature solution*, which is indeed consistent with boundary conditions and the *final steady state* temperature distribution *after a long time has passed.*

This is therefore *part* of the solution, but *there must be another solution* that will describe the interim time where the bar is cooling to this final *steady state* solution.

We next consider the final option for the separation constant applied to the spatial component:

Evaluate $\mu = -\lambda^2$.

Therefore, we obtain $\frac{u''}{u} = -\lambda^2$ and $u'' + \lambda^2 u = 0$ leads to $u(x) = A_3 \sin(\lambda x) + A_4 \cos(\lambda x)$.

So, this value of the separation constant leads to trigonometric eigenfunctions, *which* can *be made to be consistent with homogeneous boundary conditions on the bar.*

Therefore, the *Axiom for Linear PDEs* applies, since this is a linear equation, solutions rendered with the homogeneous boundary conditions can be added to other solutions to form the complete solution, so we cannot discount this possibility. Noting the superposition principle applies, if we apply homogeneous boundary conditions

$$u(0) = 0 \quad \text{and} \quad u(L) = 0$$

to the trigonometric solution, we obtain the eigenfunctions

$$u_2(x) = \sum_{n=1}^{\infty} A_n \sin\left(\frac{n \pi x}{L}\right)$$

Then, if we consider that the spatial solution is the *sum of the two solutions* that consider the true boundary conditions as well as the homogeneous boundary conditions, we have a spatially dependent solution that does indeed satisfy the boundary conditions:

$$u(x) = u_1(x) + u_2(x) = \frac{T_0 x}{L} + \sum_{n=1}^{\infty} A_n \sin\left(\frac{n \pi x}{L}\right)$$

Where $u(0) = 0$ and $u(L) = T_0$

Now we shall consider the time component of the separation constant.

So, considering the time component, we evaluate $\mu = 0$; this leads directly to

$$\mu = 0 = \frac{1}{\alpha} \frac{v'}{v} \quad \text{we obtain} \quad v_1(t) = C_1$$

Since $T(x,t) = u(x)v(t)$ then $u_1(x)v_1(t) = \frac{T_0 x}{L} C_1$

and we write that $C_1 \to 1$ to satisfy the boundary conditions. Next, we evaluate $\mu = -\lambda^2$ for the time component. Therefore,

$$\mu = -\lambda^2 = \frac{1}{\alpha} \frac{v'}{v} \quad \text{and we obtain, recalling that} \quad \lambda \to \lambda_n = \frac{n\pi}{L}$$

$$v_2(t) = C_2 \exp\left(-\alpha \frac{n^2 \pi^2}{L^2} t\right)$$

It is important to note that this is consistent with the dimensional requirements of the variables, where the argument inside the exponential must be dimensionless, giving rise to the need for the constants in the exponential, when multiplied, to result in inverse time. This is indeed the case, since from inspection:

$$\alpha \frac{n^2 \pi^2}{L^2} = \frac{k}{\rho c_p} \frac{n^2 \pi^2}{L^2} \to \frac{(W \to J/s)/(m\,K)}{(kg/m^3)(J/(kg\,K))} \frac{1}{m^2} \to \frac{1}{s}$$

Finally, again recalling that

$$T(x,t) = u(x)v(t)$$

$$T(x,t) = \frac{T_0 x}{L} + \sum_{n=1}^{\infty} C_n \sin\left(\frac{n\pi x}{L}\right) \exp\left(-\alpha \frac{n^2 \pi^2}{L^2} t\right)$$

We now apply the initial condition, where over the entire bar length L the temperature is initially fixed at $2T_0$.

$$2T_0 = \frac{T_0 x}{L} + \sum_{n=1}^{\infty} C_n \sin\left(\frac{n\pi x}{L}\right) \exp\left(-\alpha \frac{n^2 \pi^2}{L^2} 0\right)$$

Which becomes

$$2T_0 - \frac{T_0 x}{L} = \sum_{n=1}^{\infty} C_n \sin\left(\frac{n\pi x}{L}\right)$$

Orthogonality will therefore give us the final constant C_n:

$$\int_0^L \left(2T_0 - \frac{T_0 x}{L}\right) \sin\left(\frac{m\pi x}{L}\right) dx = \int_0^L \sum_{n=1}^{\infty} C_n \sin\left(\frac{n\pi x}{L}\right) \sin\left(\frac{m\pi x}{L}\right) dx$$

This results in

$$\int_0^L \left(2T_0 - \frac{T_0 x}{L}\right) \sin\left(\frac{m\pi x}{L}\right) dx = C_n \frac{L}{2}$$

Completing the integration and solving for the constant gives us

$$C_n = \frac{4T_0}{n\pi} - \frac{2T_0 \cos(n\pi)}{n\pi} = T_0\left(\frac{4 - 2\cos(n\pi)}{n\pi}\right)$$

The final solution to the time dependent heat conduction equation is then

$$T(x,t) = \frac{T_0 x}{L} + \sum_{n=1}^{\infty} T_0\left(\frac{4 - 2\cos(n\pi)}{n\pi}\right) \sin\left(\frac{n\pi x}{L}\right) \exp\left(-\alpha \frac{n^2 \pi^2}{L^2} t\right)$$

For $L = 1m$ for an aluminum bar (thermal diffusivity 97.1E-06 m²/s, $T_0 = 100°C$. This solution, using the described initial and boundary conditions, yields the following snapshots in time (0, 400, and 3600 seconds):

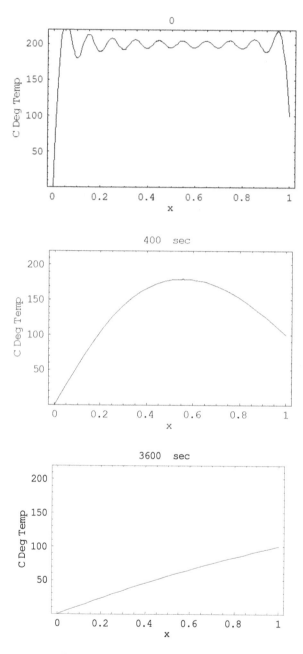

Time dependent heat solutions in bar

Method of Eigenfunction Expansion

Up to this point, we have solved the heat equation both as a steady state and a time dependent PDE. Again we will consider the solution to the time dependent heat conduction equation:

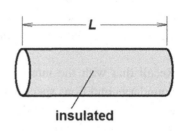

insulated

$$\nabla^2 T + \frac{q'''}{k} = \frac{1}{\alpha}\frac{\partial T}{\partial t}$$

When nuclear fission is occurring in the bar, there will be an internal heat generation term, or it could be electrically heated; either way, the internal heat generation term can vary along the length; also, heat can be removed radially along the length of the bar, or axially out of the ends; this of course depends upon the boundary conditions. To simplify our discussion, we will again consider a bar where heat is only conducted axially out of the two ends, so that the bar is again insulated along its length to prevent radial heat transfer. Now with internal heat generation, but again considering only axial heat conduction, the time-dependent heat equation becomes

$$\frac{\partial^2 T}{\partial x^2} + \frac{q'''}{k} = \frac{1}{\alpha}\frac{\partial T}{\partial t}$$

We will also consider *Initial* and *Boundary conditions* as follows:

1) $T(x,0) = T_0$ (Initial Condition)
2) $T(0,t) = 0$
3) $T(L,t) = 0$

Normally, at this point we would usually try to apply *Separation of Variables* to solve this equation. However, try as we might, this time dependent heat equation cannot be readily separated when we have a non-zero internal heat generation term. So, it is useful to point out that in solving all differential equations, the method of solution typically begins by analyzing the equation, and trying a solution method based on analysis and intuition (which comes from practice!). The method of *Eigenfunction Expansion* is like this also: it involves looking at a PDE and considering how we might leverage orthogonality with eigenfunctions that solve a simpler, similar, yet related PDE.

Recall that with the internal heat generation term set to zero, with the boundary conditions specified, we would obtain the eigenfunctions as we did in the previous time dependent heat conduction problem (following the procedure of investigating all of the separation constant possibilities for the given *and* homogeneous boundary condition cases), so that the eigenfunctions we obtained were of the form

$$\sum_{n=1}^{\infty} A_n \sin\left(\frac{n \pi x}{L}\right)$$

Since these eigenfunctions solved the time dependent heat equation without the internal heat generation term, Eigenfunction Expansion will be applied using these same eigenfunctions as the *assumed form for the spatial component* of the equation, which can be leveraged using orthogonality to match the conditions that we need it to and thus give us the solution to the PDE. Therefore, we will assume that the solution to this problem is

$$T(x,t) = \sum_{n=1}^{\infty} U_n(t) \sin(\frac{n \pi x}{L})$$

If this is true, then we can write

$$\frac{\partial^2 T}{\partial x^2} = \sum_{n=1}^{\infty} -U_n(t) \left(\frac{n \pi}{L}\right)^2 \sin(\frac{n \pi x}{L})$$

$$\frac{\partial T}{\partial t} = \sum_{n=1}^{\infty} \frac{\partial U_n(t)}{\partial t} \sin(\frac{n \pi x}{L})$$

Next, we reorder the original governing partial differential equation:

$$\frac{\partial T}{\partial t} - \alpha \frac{\partial^2 T}{\partial x^2} = \alpha \frac{q'''}{k}$$

One important issue is that we must have the assumed form of the eigenfunctions available as a basis for each term in the equation. Therefore, we assume that we can suitably expand the internal heat generation term using the eigenfunctions

$$f(x,t) = \alpha \frac{q'''}{k} = \sum_{n=1}^{\infty} f_n(t) \, \sin(\frac{n \pi x}{L})$$

To determine $f_n(t)$ we apply orthogonality. Therefore, to yield the non-zero term, we note that $m = n$, and therefore we will drop the sum in the following expression:

$$\int_0^L \left(\alpha \frac{q'''}{k} \right) \sin\left(\frac{m \pi x}{L} \right) dx = \int_0^L \sum_{n=1}^{\infty} f_n(t) \sin\left(\frac{n \pi x}{L} \right) \sin\left(\frac{m \pi x}{L} \right) dx$$

Simplifying this results in:

$$f_n(t) = \frac{2 \alpha q'''}{k n \pi} (1 - \cos(n\pi))$$

Then, reassembling the differential equation using each term based on the same eigenfunctions, we obtain

$$\sum_{n=1}^{\infty} \left(\frac{\partial U_n(t)}{\partial t} + \alpha \left(\frac{n \pi}{L} \right)^2 U_n(t) = \frac{2 \alpha q'''}{k n \pi} (1 - \cos(n\pi)) \right) \sin(\frac{n \pi x}{L})$$

Therefore, the next step is to recognize that we must solve the first order time dependent equation

$$\frac{\partial U_n(t)}{\partial t} + \alpha \left(\frac{n \pi}{L} \right)^2 U_n(t) = \frac{2 \alpha q'''}{k n \pi} (1 - \cos(n\pi))$$

To simplify this equation, we make the following substitutions

$$\lambda_n = \alpha \left(\frac{n \pi}{L} \right)^2 \quad \text{and} \quad \mu_n = \frac{2 \alpha q'''}{k n \pi} (1 - \cos(n\pi))$$

We therefore have the first order equation

$$\frac{\partial U_n(t)}{\partial t} + \lambda_n U_n(t) = \mu_n$$

This can be solved by the method of integrating factor, which is

$$\exp(\int \lambda_n dt) = e^{\lambda_n t}$$

This leads to

$$\frac{dU_n(t)}{dt} e^{\lambda_n t} + \lambda_n e^{\lambda_n t} U_n(t) = \mu_n e^{\lambda_n t}$$

Which then becomes, integrating both sides

$$\int d(U_n(t)e^{\lambda_n t}) = \int \mu_n e^{\lambda_n t} dt$$

This results in

$$U_n(t) = \frac{\mu_n}{\lambda_n} + c_n e^{-\lambda_n t}$$

Assembling the complete solution, we obtain

$$T(x,t) = \sum_{n=1}^{\infty} \left(\frac{\mu_n}{\lambda_n} + c_n e^{-\lambda_n t} \right) \sin(\frac{n \pi x}{L})$$

This satisfies the given boundary conditions and solves the differential equation. Now, we have to determine the constant, which comes from the remaining initial condition. This results in

$$T(x,0) = T_o = \sum_{n=1}^{\infty} \left(\frac{\mu_n}{\lambda_n} + c_n e^{-\lambda_n \cdot 0} \right) \sin(\frac{n \pi x}{L})$$

We can then apply orthogonality to isolate the nth eigenvalue constant

$$\int T_o \sin(\frac{n\pi x}{L})dx = \int \sum_{n=1}^{\infty} \left(\frac{\mu_n}{\lambda_n} + c_n \right) \sin^2(\frac{n\pi x}{L})dx$$

Performing the integration and simplifying, we obtain

$$c_n = \left(\frac{2T_o}{n\pi}(1 - \cos(n\pi)) - \frac{\mu_n}{\lambda_n} \right)$$

Therefore, the solution becomes

$$T(x,t) = \sum_{n=1}^{\infty} \left(\frac{\mu_n}{\lambda_n} + \left(\frac{2T_o}{n\pi}(1 - \cos(n\pi)) - \frac{\mu_n}{\lambda_n} \right) e^{-\lambda_n t} \right) \sin(\frac{n\pi x}{L})$$

Recalling that

$$\lambda_n = \alpha \left(\frac{n\pi}{L} \right)^2 \qquad \text{and} \qquad \mu_n = \frac{2\alpha q'''}{k n\pi}(1 - \cos(n\pi))$$

$$\text{in SI units of } 1/s \qquad\qquad \text{in SI units of } °K/s$$

And simplifying, we obtain the final and complete solution using the method of eigenfunction expansion

$$T(x,t) =$$

$$\sum_{n=1}^{\infty} \left(\frac{\frac{2q'''}{k n\pi}(1 - \cos(n\pi))}{\left(\frac{n\pi}{L} \right)^2} (1 - e^{-\alpha \left(\frac{n\pi}{L} \right)^2 t}) + \left(\frac{2T_o}{n\pi}(1 - \cos(n\pi)) \right) e^{-\alpha \left(\frac{n\pi}{L} \right)^2 t} \right) \sin(\frac{n\pi x}{L})$$

This solution is difficult to envision just by inspection, but rest assured, it produces a result that indeed makes physical sense. Let us suppose that the bar is a "slug" of spent nuclear fuel, and the internal heat generation therefore comes from fission product decay from irradiation in the reactor, assumed to be at a rate of 10 W/m³. Note this decay heat generation would depend on the reactor conditions, "burnup," and isotopic content at discharge—determined among other things from application of Bateman's equations on the isotopic production, which is not trivial.

The internal heat generation would also not remain constant, and the rate of heat generation would decay with some effective overall exponential decay constant; however, let us assume that for our purposes here, it is constant over the time period considered. If the bar is wrapped radially with insulation, then heat transfer is by conduction only out of the ends. Furthermore, assume the bar is initially at 130C, is 0.10 m in length, a thermal diffusivity of $\alpha = 1.1811E-5$ m2/s, and a conductivity $k = 27$ W/(m K), values reasonable for uranium metal.

Finally, the boundary conditions are such that the ends are suddenly held at 100C, so that our eigenfunction expansion solution to the heat equation directly applies. *To create homogeneous boundary conditions on the ends, we can subtract 100C from all conditions. In the end we can add this back in.* Plotting the solution (through $n = 25$ terms), we obtain a very physical result, as shown in the figure below.

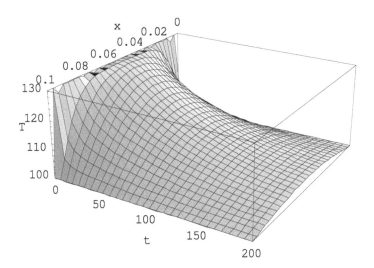

Plot of temperature in a uranium nuclear fuel slug discharged from a reactor

Inspection of the figure reveals that the slug is indeed at a uniform temperature of 130C initially, albeit with the expected Gibbs' phenomenon evident neat the edges. As the ends are suddenly cooled, the evidence of heat removal out of the ends is clear and quite physical.

Chapter 12

Applied Solution Methods—Part 3: Neutronics

We now consider problems involving neutron transport and neutron diffusion theory; because the emphasis of this presentation is on *methods*, we do not present strict end-to-end theory to support/not support the use of diffusion theory in the applied problems considered, as this is best covered in a dedicated course on reactor physics. We begin with the transport equation for completeness, and then merge into the diffusion equation.

Overall, we are interested in the determination of the neutron distribution in the reactor system, which is directly related to the fission power, and tracking the neutron population, since these are the "fission chain carrier" in the reactor. The Neutron Transport Equation describes the precise behavior of neutrons in the reactor system. In some cases, the neutron diffusion equation offers an approximation to the transport equation, except in regions where flux gradients are steep, e.g. at material interfaces with strong absorbers, or near boundaries; in these instances, the differences between transport and diffusion solutions can be 50% or more.

The Neutron Transport Equation

The linear neutral particle radiation transport equation was developed by Boltzmann in the 19th Century to track the flow of neutral particles as a function of *angle, energy,* and *space,* far ahead of the Manhattan Project. In any case, the neutral particles we are concerned about in nuclear systems are mainly neutrons or photons, principally associated with fission in fuel to determine neutron balance in multiplying chain reacting systems, and we assume they interact in a mass bounded by a closed surface.

Mathematically, the transport equation offers a simple balance, where it states that particles can be lost by leakage or collision, but then can be gained by scatter back into the system or produced by fission or an independent source. Discretization of particle energy is accomplished by spectrally averaging over energy groups ($g = 1, G$), from high to low energies, resulting in the so called "multigroup transport" formulation. In steady state, the multigroup transport equation is as follows, where the left side includes loss by leakage and collision, respectively, and scatter from other (prime) directions and energies scatter into the energy and angle $\hat{\Omega}$ of interest, along with fission production density and independent sources on the right:

$$\hat{\Omega} \cdot \nabla \psi_g(\vec{r}, \hat{\Omega}) + \sigma_g(\vec{r}) \psi_g(\vec{r}, \hat{\Omega}) =$$

$$\sum_{g'=1}^{g} \int_{4\pi} d\Omega' \sigma_{s\,g' \to g}(\vec{r}, \hat{\Omega}' \cdot \hat{\Omega}) \psi_g(\vec{r}, \hat{\Omega}') + \frac{\chi_g}{k_o} \sum_{g'=1}^{G} \int_{4\pi} d\Omega' \nu \sigma_{f\,g'}(\vec{r}) \psi_{g'}(\vec{r}, \hat{\Omega}) + q_{ind\,g}(\vec{r}, \hat{\Omega})$$

The units of each summed term are n/(cm^3 s sr).

The Angular Variable and Legendre Moments

Note that the angular variable is normalized on the unit sphere in the above formulation, so that integration over Ω is expressed in terms of the polar angle cosine μ and azimuthal angle φ as:

$$\int_{4\pi} d\Omega = \int_{-1}^{1} \frac{d\mu}{2} \int_{0}^{2\pi} \frac{d\varphi}{2\pi} = 1$$

Hereafter, this is implicitly assumed. The scattering term is then expanded using a truncated set of spherical (surface) harmonics, with

$$\hat{\Omega} \to <\theta, \varphi>, \quad (\hat{\Omega}' \cdot \hat{\Omega}) \to (\mu_o), \quad \mu = (\cos \theta), \quad \mu' = (\cos \theta')$$

Where the scatter cross section is expanded in Legendre moments (as discussed previously):

$$\sigma_{s\,g' \to g}(\vec{r}, \mu_o) = \sum_{l=0}^{L} (2l+1) \sigma_{s\,g' \to g,l}(\vec{r}) P_l(\mu_o)$$

$$\sigma_{s\,g'\rightarrow g,l}(\vec{r}) = \int_{-1}^{1} \frac{d\mu_o}{2} \sigma_{s\,g'\rightarrow g}(\vec{r},\mu_o) P_l(\mu_o)$$

$$\mu_o = \mu\mu' + (1-\mu^2)^{1/2}(1-\mu'^2)^{1/2}\cos(\varphi-\varphi')$$

The Legendre polynomial $P_l(\mu_o)$, using the Legendre Addition Theorem, is defined using the spherical harmonics functions:

$$P_l(\mu_o) = \frac{1}{(2l+1)} \sum_{k=-l}^{l} Y_{l,k}^*(\theta',\varphi') Y_{l,k}(\theta,\varphi)$$

The spherical (surface) harmonics $Y_{l,k}$ and $Y_{l,k}^*$ are defined in terms of the Associated Legendre polynomials and an exponential term:

$$Y_{l,k}(\theta,\varphi) = \sqrt{(2l+1)\frac{(l-k)!}{(l+k)!}} P_l^k(\mu) \exp(ik\varphi)$$

$$Y_{l,-k}(\theta,\varphi) = (-1)^k Y_{l,k}^*(\theta,\varphi)$$

Using previous equations, $P_l(\mu_o)$ can be written:

$$P_l(\mu_o) = P_l(\mu)P_l(\mu') + 2\sum_{k=1}^{l} \frac{(l-k)!}{(l+k)!} P_l^k(\mu)P_l^k(\mu')\cos(k(\varphi-\varphi'))$$

By trigonometric identity:

$$\cos(k(\varphi-\varphi')) = \cos(k\varphi)\cos(k\varphi') + \sin(k\varphi)\sin(k\varphi')$$

The vectors $\hat{\Omega} \rightarrow <\theta,\varphi>$ on the unit sphere can be expressed as a set of direction cosines projected parallel to the x, y, and z axes, respectively, as $<\mu,\eta,\xi>$ Note that η and ξ can be expressed in terms of polar angle cosine μ and the azimuthal angle φ:

$$\eta = \sqrt{1-\mu^2}\cos(\varphi) \qquad \xi = \sqrt{1-\mu^2}\sin(\varphi)$$

A 3-D Cartesian geometry (using a right handed coordinate system) is shown in Figure below.

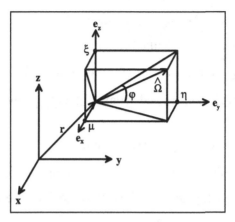

3-D Cartesian Geometry transport coordinate system for a particle in group g at $(r, \hat{\Omega})$

If the streaming operator $\hat{\Omega} \cdot \nabla$ is expanded in 3-D Cartesian coordinates, it becomes:

$$\hat{\Omega} \cdot \nabla = \mu \frac{\partial}{\partial x} + \eta \frac{\partial}{\partial y} + \xi \frac{\partial}{\partial z}$$

3-D Cartesian Boltzmann Transport Equation

Substituting equations back into the transport equation, we obtain the Legendre expanded multigroup form of the transport equation in 3-D Cartesian geometry considering only fission sources:

$$(\mu \frac{\partial}{\partial x} + \eta \frac{\partial}{\partial y} + \xi \frac{\partial}{\partial z}) \psi_g (x, y, z, \mu, \varphi) + \sigma_g (x, y, z) \psi_g (x, y, z, \mu, \varphi) =$$

$$\sum_{g'=1}^{G} \sum_{l=0}^{L} (2l+1) \sigma_{s, g' \to g, l} (x, y, z) \{ P_l (\mu) \phi_{g', l} (x, y, z) + 2 \sum_{k=1}^{l} \frac{(l-k)!}{(l+k)!} P_l^k (\mu) \cdot$$

$$[\phi_{C g', l}^k (x, y, z) \cos(k\varphi) + \phi_{S g', l}^k (x, y, z) \sin(k\varphi)] \} + \frac{\chi_g}{k_o} \sum_{g'=1}^{G} \nu \sigma_{f g'} (x, y, z) \phi_{g', 0} (x, y, z)$$

where:

$\mu = x$ direction cosine for angular ordinate

$\eta = y$ direction cosine for angular ordinate

$\xi = z$ direction cosine for angular ordinate

ψ_g = group g angular particle flux (for groups $g=1,G$)

φ = azimuthal angle constructed from $\arctan(\xi/\eta)$, with proper phase shift

σ_g = total group macroscopic cross section

l = Legendre expansion index ($l = 0, L$), $L = 0$ or odd truncation

$\sigma_{sg'\to g,l}$ = l th Legendre moment of the macroscopic differential scattering cross section from group $g' \to g$

$P_l(\mu) = l$ th Legendre polynomial

$\phi_{g',l} = l$ th Legendre scalar flux moment for group g

$P_l^k(\mu) = l$ th, k th Associated Legendre polynomial

$\phi_{C\,g',l}^k = l$ th, k th Cosine Associated Legendre flux moment, group g

$\phi_{S\,g',l}^k = l$ th, k th Sine Associated Legendre scalar flux moment, group g

χ_g = group fission distribution constant (neutrons)

k_o = criticality eigenvalue (neutrons)

$\nu\sigma_{f\,g}$ = group fission production (neutrons)

The flux moments given below are defined in terms of μ' and φ' as:

$$\phi_{g',l}(x,y,z) = \int_{-1}^{1} \frac{d\mu'}{2} P_l(\mu') \int_{0}^{2\pi} \frac{d\varphi'}{2\pi} \psi_{g'}(x,y,z,\mu',\varphi')$$

$$\phi^k_{C\,g',l}(x,y,z) = \int_{-1}^{1} \frac{d\mu'}{2} P_l^k(\mu') \int_{0}^{2\pi} \frac{d\varphi'}{2\pi} \cos(k\varphi') \psi_{g'}(x,y,z,\mu',\varphi')$$

$$\phi^k_{S\,g',l}(x,y,z) = \int_{-1}^{1} \frac{d\mu'}{2} P_l^k(\mu') \int_{0}^{2\pi} \frac{d\varphi'}{2\pi} \sin(k\varphi') \psi_{g'}(x,y,z,\mu',\varphi')$$

Solution of the transport equation can proceed in a variety of ways; typically, it is solved deterministically using the discrete ordinates (Sn) method; this is where a number of (usually symmetric) angles are considered on the unit sphere, and the transport equation is solved along each angle.

Forward Transport Boundary Conditions

Boundary conditions for transport problems are typically either "Reflective" (Specular Reflective), Albedo (partial reflection a returned fraction of incident flux), or Vacuum, where angular fluxes at a vacuum surface interface "S" have zero values for directions into the solid, as indicated in the figure below

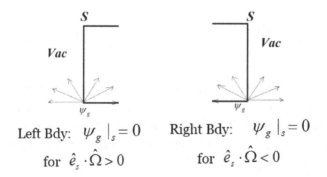

Left Bdy: $\psi_g\,|_s = 0$ Right Bdy: $\psi_g\,|_s = 0$

for $\hat{e}_s \cdot \hat{\Omega} > 0$ for $\hat{e}_s \cdot \hat{\Omega} < 0$

Boundary conditions for transport; a Vacuum boundary (Bdy) with a surface S interface is where particles along any direction crossing S leaking into the vacuum cannot return back into the solid

The scalar flux is the zeroth moment in the flux moment equations comes from an integration (quadrature formula) applied to the energy group dependent angular flux over all directions of the solid angle. In one dimensional deterministic transport, Gaussian quadrature is used, where the ordinates selected for the collection of angular fluxes are the direction cosines from the roots of the Legendre Polynomials.

Adjoint Radiation Transport

The adjoint transport operator H^+ can be derived using the *adjoint identity* for real valued functions and the forward multi-group transport operator, where $\langle \ \rangle$ represents integration over all independent variables:

$$\langle \psi_g^+ \, H \, \psi_g \rangle = \langle \psi_g \, H^+ \psi_g^+ \rangle$$

$$H = \hat{\Omega} \cdot \nabla + \sigma_g(\vec{r}) - \sum_{g'=1}^{G} \int_{4\pi} d\Omega' \, \sigma_{s\,g'\to g} \, (\vec{r}, \hat{\Omega}' \cdot \hat{\Omega})$$

The angular adjoint (importance) function is ψ_g, and H is the forward transport operator. Applying the adjoint boundary condition that particles leaving a bounded system have an importance of zero in all groups (converse of the vacuum boundary condition in a forward calculation) with the above equations, and requiring a continuous importance function mathematically leads to the multi-group adjoint transport operator:

$$H^+ = -\hat{\Omega} \cdot \nabla + \sigma_g(\vec{r}) - \sum_{g'=1}^{G} \int_{4\pi} d\Omega' \, \sigma_{s\,g\to g'} \, (\vec{r}, \hat{\Omega} \cdot \hat{\Omega}')$$

This operator is not Hermitian, and has units of inverse length. Also, the minus sign on the streaming term is an indicator that adjoint particles travel along a reversed direction, where scattering progresses from group *g back* to other groups *g'* (those groups formerly contributing to group *g* in the forward transport operator).

Note that the adjoint transport operator, H^+ can also be derived directly from physical principles based on the conservation of neutron importance.

This type of derivation does not require a strict mathematical application of vector identities and a *zero importance* condition for particles leaving the system. No matter how it is derived, the adjoint function is associated with the importance of neutral particles with respect to some objective. For example, the particles can be neutrons and the objective could be an absorption in a He-3 neutron detector to yield an (n,p) reaction.

To solve for the adjoint function, a forward transport solver can be directly used if angular, cross section, and energy group indices are properly treated. That is, a forward transport algorithm can be used to solve an adjoint transport problem if the group cross sections and sources are transposed, including the cross section scattering matrix, re-ordered to commence from group G to 1. In this case, all angles are considered to be defined implicitly in opposite (negative) directions, with group G adjoint sources input and reported into the forward code as group 1, and group *G-1* adjoint sources input and reported as group 2, etc.

Adjoint Boundary Conditions

Adjoint Boundary Conditions can be reflective, albedo, or vacuum, but are different since the adjoint function is related to the efficiency for a particular response. The Vacuum Addjoint conditions specify that particles leaving a bounded surface have an importance function of zero.

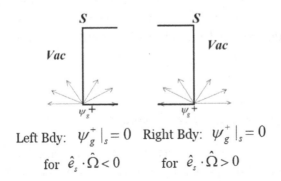

Left Bdy: $\psi_g^+ \big|_s = 0$ Right Bdy: $\psi_g^+ \big|_s = 0$

for $\hat{e}_s \cdot \hat{\Omega} < 0$ for $\hat{e}_s \cdot \hat{\Omega} > 0$

Boundary conditions for adjoint transport; particles leaving the system crossing S along any direction are assigned a "zero" adjoint importance (efficiency)

Once solved for, the dimensionless adjoint function provides the neutron or photon importance throughout the problem phase space relative to a particular response, defined by the adjoint source. The adjoint can then be used in several ways. Ultimately, we wish to predict the overall response R of a detector in counts per second using the adjoint function. For problems of this type, the count rate R is based on the number of absorption reactions per unit time in the He-3 gas (leading to an (n,p) reaction) caused by neutrons. If such a fixed neutron source/detector problem is proposed, the neutron flux must satisfy the transport equation:

$$H\psi_g = q_g$$

and the inhomogeneous adjoint equation should be satisfied with an adjoint source aliased to the group detector response cross section σ_{dg}, according to:

$$H^+\psi_g^+ = \sigma_{dg}$$

Applying applicable formulations to the adjoint identity equation and integrating over all variables results in a very useful expression for detector response R:

$$R = \left\langle \psi_g \sigma_{dg} \right\rangle = \left\langle \psi_g^+ q_g \right\rangle$$

This indicates that the detector response can be obtained by complete integration of the source distribution with the adjoint function obtained from the adjoint transport equation. Therefore, note that R can be computed directly from the results of either of (i) *several* forward transport computations for each neutron source, or (ii) a single adjoint transport computation.

Forward Radiation Transport Response

In reference to a He-3 detector, in the traditional forward case, the standard Boltzmann transport equation must be solved to yield a scalar flux for a specific neutron source q placed inside the Sample Chamber. For a number of different types of sources, *separate computations* must be performed for *each source*. Once the neutron flux is determined for each scenario, the detector response in the He-3 could be numerically computed in the conventional manner using group cell fluxes and detector cross sections, as given below:

$$R = \int_{V_d, \forall E} \phi(x, y, z, E)\, \sigma_d(x, y, z, E)\, dx\, dy\, dz\, dE \;\approx\; \sum_{\substack{\Delta V_i \in V_d \\ g=1,G}} \phi_{g,i}\, \sigma_{d\,g,i}\, \Delta V_i$$

where: R = Detector Response, counts/s

V_d = Volume occupied by the detector, in cm^3

(x, y, z) = Spatial Location(s) of He-3 Tubes

$\phi(x, y, z, E)$ = Spatial, Energy Dependent Scalar Flux, n/cm^2/s

$\sigma_d(x, y, z, E)$ = Spatial, Energy Dependent Detector Cross Section, cm^2/cm^3

$\phi_{g,i}$ = ith Cell Scalar Flux for Group g, n/cm^2/s, from quadrature of $\psi_{g,i}$

$\sigma_{d\,g,i}$ = ith Cell Detector Cross Section for Group g, cm^2/cm^3

ΔV_i = ith Cell Volume, cm^3

Adjoint Transport Response

In the adjoint case, the adjoint transport equation must be solved using an *adjoint source* that is equal in magnitude to the detector response cross section placed in each location, for this discussion, occupied by He-3 tubes. This yields the adjoint function throughout the problem phase space, and represents the importance of neutrons in each spatial location and energy group relative to a response in the collective of He-3 tubes. Then, the He-3 tube count rate R due to *any* neutron source q placed in a specified location is computed as given below:

$$R = \int_{V_q, \forall E} \phi^{+}{}_d(x', y', z', E)\, q(x', y', z', E)\, dx'\, dy'\, dz'\, dE \;\approx\; \sum_{\substack{\Delta V_i \in V_q \\ g=1,G}} \phi^{+}{}_{d\,g,i}\, q_{g,i}\, \Delta V_i$$

where:

R = Detector Response, counts/s

V_q = Source Volume, cm^3 (volume occupied by the source)

(x', y', z') = Spatial Location of Nonzero Source Cells

$\phi^+_d(x', y', z', E)$ = Scalar Adjoint Function for Detector d, from quadrature

$q\,(x', y', z', E)$ = Spatial, Energy Dependent Source, $n/cm^3/s$

$\phi^+_{d\,g,i}$ = ith Cell Scalar Adjoint Function for Group g, Detector d

$q_{g,i}$ = ith Cell Source Density for Group g, $n/cm^3/s$

ΔV_i = ith Cell Volume, cm^3

Therefore, the appropriate adjoint source (as discussed for this application) is a unit source weighted by the group absorption cross sections for He-3, placed in each He-3 location. The phase space integral of any arbitrary cell source distribution weighted by the computed adjoint function yields the total detector response R. Hence, a single adjoint calculation may be used to predict R from any conceivable source distribution.

One should make note of the duality of the response R predicted by theory in and stated in the equations above. However, responses computed using forward and adjoint solutions can only be directly comparable if the numerical truncation error from either computation is negligible. In general, responses are typically comparable within some prescribed margin of error as a result of numerical effects.

NUCLEAR APPLICATION: Adjoint Importance of neutrons

Once solved for, the adjoint function provides the neutron or photon importance throughout the problem phase space relative to a particular response, defined by the adjoint source. The adjoint function can then be used in several ways.

One use of the adjoint function is to determine the regions/energies that most affect the response to help determine limiting mesh intervals in the geometry. Further, a deterministic adjoint solution may be used to assign importances for variance reduction in non-analog Monte-Carlo applications, which can add extreme efficiency to such calculations.

Neutron Diffusion Equation

How neutrons scatter and interact in materials can be approximated under certain conditions by the neutron diffusion equation. To understand this equation, we need to derive it term by term. From now on, for illustrations regarding neutron transport and diffusion, we will only consider monoenergetic neutrons. We must next introduce the concept of a *reaction rate density*, which is the number of neutron reactions per cubic centimeter per second. The reaction rate density is the product of macroscopic cross section and the incident particle scalar flux.

Scalar flux, often represented by ϕ, is defined as the product of the neutron density (the number of neutrons per cubic centimeter at a particular location) and the speed of the neutrons.

$$\phi\left(\frac{n}{cm^2 s}\right) = n\left(\frac{n}{cm^3}\right) v\left(\frac{cm}{s}\right)$$

To simplify the discussion, have assumed that neutrons all have the same energy, and therefore all have the same speed. In reality, the energy, and therefore the speed of neutrons is a probability distribution function, governed by the physics of the problem under consideration; for example, consider the fission neutron distribution function. The more complex details of how to handle these distributions can be addressed in a dedicated study of reactor physics.

For now, it should be noted that one definition of the scalar flux is that it is the total path length traveled by all particles in any direction contained in a volume, divided by that volume, in one second. (We further note that this is precisely how "volumetric flux tallies" are computed in Monte Carlo codes that track neutrons or gamma rays using the Monte Carlo method for determining particle interactions)

Now consider an arbitrary volume V composed of a homogeneous material.

For the material, the neutron absorption cross section Σ_a represents the cross sectional area of a nucleus presented to a neutron per unit volume:

$$\Sigma_a \quad \frac{cm^2}{cm^3} \rightarrow cm^{-1}$$

When this is multiplied by the neutron flux, this quantity is a reaction rate for neutron absorption, R_a, which is the number of neutrons absorbed per unit volume per unit time:

$$R_a \left(\frac{\#}{cm^3 s} \right) = \Sigma_a \left(\frac{1}{cm} \right) \phi \left(\frac{n}{cm^2 s} \right)$$

Then the total number of neutron absorption reactions occurring in V occurring per unit time are:

$$\iiint_V \Sigma_a \phi \, dV$$

The total leaking out of V, relative to outward pointing differential surface vectors, is:

$$\iint_S \bar{J} \cdot d\bar{s}$$

The total neutrons produced inside volume V:

$$\iiint_V S_o\, dV$$

Where S_o is the neutron source production density, with units of $n/(cm^3\, s)$. The neutrons in volume V must balance if there is a steady state: those leaking out and that are absorbed must be precisely balanced by the amount of source neutrons produced:

$$\iint_S \bar{J} \cdot d\bar{s} + \iiint_V \Sigma_a \phi\, dV = \iiint_V S_o\, dV$$

Gauss' Divergence Theorem equates a surface integral and a volume integral:

$$\iint_S \bar{J} \cdot d\bar{s} = \iiint_V \nabla \cdot \bar{J}\, dV$$

This enables us to put all terms in the equation according to a volume integral. Therefore, using the Divergence Theorem and simplifying:

$$\iiint_V \left(-\nabla \cdot \bar{J} - \Sigma_a \phi + S_o \right) dV = 0$$

Therefore, on a differential basis we can drop the integral, assuming the balance must still hold on a differential basis, subtracting leakage and absorption and adding source, again assuming steady state, we obtain the neutron balance equation:

$$-\nabla \cdot \bar{J} - \Sigma_a \phi + S_o = 0$$

If one integrates the one-speed neutron transport equation (presented earlier in this chapter) over all angles (ordinates), this same "neutron balance equation" is the result.

If not in steady state, then the balance equation describes the time rate of change of the neutron density, which can be equated by the time rate of change in flux divided by the speed:

$$-\nabla \cdot \vec{J} - \Sigma_a \phi + S_o = \frac{\partial n}{\partial t} = \frac{1}{v}\frac{\partial \phi}{\partial t}$$

The neutron current $\vec{J}$ refers to neutrons traveling in a particular direction. In neutron diffusion theory, it is assumed that the link between the current $\vec{J}$ and flux ϕ is described by Fick's Law:

$$\vec{J} = -D\nabla \phi$$

Where the diffusion constant, derived from the P_1 approximation to the transport equation, is equal one third of the transport (tr) mean free path, stated here without proof:

$$D = \frac{1}{3\Sigma_{tr}} = \frac{1}{3(\Sigma_t - \bar{\mu}_o \Sigma_s)}$$

The transport cross section is the total cross section minus the scatter cross section weighted by the average cosine of the scattering angle $\bar{\mu}_o = 2/(3A)$ for scattering with a nucleus of mass number A for scatter that is isotropic in the "Center of Mass" scattering system (as covered in reactor physics).

Fick's law therefore states that to first order the neutron current is proportional (by a constant of proportionality D) to the gradient of the flux. The units of the current are the same as those for flux, e.g. $n/(cm^2\ s)$; however, the current refers to net neutron *flow* relative to a particular vector direction. For completeness, we note that this same relationship between flux and current given by Fick's law can be derived using the "P_1 Approximation" to the Boltzmann Transport Equation, which assumes that the angular flux ψ is truncated to linearly dependence on Ω.

This is discussed in detail in a course on nuclear reactor physics; while of great interest, the theory is beyond the scope of the discussion here, where we focus on the mechanics of solving the equations of neutron balance. To clarify further the concept of neutron Current or net flow of neutrons through a normal surface area, recall that in Cartesian coordinates, the Del operator is, using vector and unit vector notation:

$$\nabla = \left\langle \frac{\partial}{\partial x}, \frac{\partial}{\partial y}, \frac{\partial}{\partial z} \right\rangle = \frac{\partial}{\partial x}\hat{i} + \frac{\partial}{\partial y}\hat{j} + \frac{\partial}{\partial z}\hat{k}$$

And using the Del operator, operating on a scalar function f is

$$grad \; f = \nabla f = \frac{\partial f}{\partial x}\hat{i} + \frac{\partial f}{\partial y}\hat{j} + \frac{\partial f}{\partial z}\hat{k}$$

Therefore, returning to the discussion of neutron current,

$$\vec{J} = -D\left(\frac{\partial \phi}{\partial x}\hat{i} + \frac{\partial \phi}{\partial y}\hat{j} + \frac{\partial \phi}{\partial z}\hat{k} \right)$$

The magnitude of the current along x, or net neutron flow (relative to a vector direction along the x-axis) is:

$$J_x = \vec{J}\cdot\hat{i} = -D\frac{\partial \phi}{\partial x}$$

Similarly, the magnitude of the net neutron flow (relative to a vector direction along the y- and z-axes) respectively:

$$J_y = \vec{J}\cdot\hat{j} = -D\frac{\partial \phi}{\partial y}$$

$$J_z = \vec{J}\cdot\hat{k} = -D\frac{\partial \phi}{\partial z}$$

Then, we apply Fick's Law to make the requisite link between leakage and flux:

$$\vec{J} = -D\nabla\phi$$

The time dependent diffusion equation becomes:

$$-\nabla \cdot (-D\nabla \phi) - \Sigma_a \phi + S_o = \frac{\partial n}{\partial t} = \frac{1}{v}\frac{\partial \phi}{\partial t}$$

With the Diffusion constant "D" as a constant, we can simplify

$$D\nabla^2 \phi - \Sigma_a \phi + S_o = \frac{\partial n}{\partial t} = \frac{1}{v}\frac{\partial \phi}{\partial t}$$

Again, referring back to the neutron transport equation, the angular dependent flux ψ can be expanded using othrogonal spherical harmonics functions. If these are truncated to first order, this results in the "P_1 approximation" to the transport equation.

While this subject is treated in detail in a text on reactor physics, we can use it here to illustrate *partial neutron currents* based on the P_1 approximation:

$$J^\pm = \int_{2\pi+} d\hat{\Omega}\hat{e}_s \cdot \hat{\Omega}\psi \Rightarrow \frac{\phi}{4} \mp \frac{D}{2}\nabla \phi \cdot \hat{e}_s$$

Stated separately:

$$J^+ = \frac{\phi}{4} - \frac{D}{2}\nabla \phi \cdot \hat{n}$$

$$J^- = \frac{\phi}{4} + \frac{D}{2}\nabla \phi \cdot \hat{n}$$

Based on these definitions, we also note that Fick's Law is preserved:

$$\left(J^+ - J^-\right) = J_{NET} = -D\nabla \phi \cdot \hat{n} = \left|\vec{J}\right|$$

Diffusion Boundary Conditions

The equivalent of a "Vacuum" boundary condition, or "Zero return current" boundary, where neutrons (particles) cannot scatter back across the surface that they exited from in a vacuum, comes from the partial current condition:

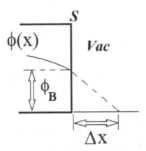

Vacuum Boundary Extrapolation distance for diffusion theory—this is a distance by which we extend the boundary to enable the correct solution at the real physical boundary

Since we set the return partial current (J^-) at surface S equal to zero, this means:

$$J^- |_s = 0 = \frac{\phi_B}{4} + \frac{D}{2} \nabla \phi \cdot \hat{e}_s = \frac{\phi_B}{4} + \frac{D}{2} \frac{\partial \phi}{\partial x} |_s$$

Then, since the boundary flux in the real boundary surface of the slab is ϕ_B, we cqn approximate the slope with a "rise/run" estimate as we extrapolate the flux to a zero value to enable us to obtain the correct value on the real physical slab boundary by a distance Δx, which we can show must equal 2D, as follows:

$$J^- = 0 \approx \frac{\phi_B}{4} + \frac{D}{2} \left(\frac{-\phi_B}{\Delta x} \right) \rightarrow \Delta x = 2D$$

Therefore, the diffusion boundary condition known as an "*extrapolated (vacuum) boundary condition*" calls for setting the flux to zero at a distance of $2D$ past the physical boundary. For example, a vacuum extrapolated boundary the right edge of a slab of dimension a is often denoted as $\tilde{a} = a + 2D$. (Note: A more exact extrapolation distance predicted by transport theory gives this correction as $\tilde{a} = a + 2.13D$, which gives slightly improved accuracy more consistent with transport theory results—this is known as a "*Transport corrected extrapolated boundary*").

NUCLEAR APPLICATION: *Steady State Critical Sphere*

A critical mass of nuclear fissile fuel is one that has neutron production from fission exactly balanced by neutron loss through absorption and neutron leakage out of the surface. Therefore, if we implicitly assume the neutron diffusion equation is valid for our discussion of determining a critical mass, then we have the equation

$$D\nabla^2 \phi - \sigma_a \phi + \frac{\nu \sigma_f \phi}{k} = 0$$

(leakage) – (absorption) + (source) = (steady state balance, net 0)

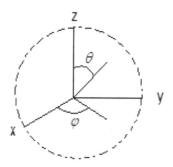

A spherical plutonium ball

If this equation indeed describes the neutron behavior of our critical mass, then the value of k (called the *criticality eigenvalue*) is precisely unity. In this example, let us consider a bare (no reflector) plutonium sphere placed in a vacuum. We must determine the size of the sphere that will achieve a critical condition. At this point, we can assume our plutonium has the following properties for the neutron diffusion constant, and absorption and fission production cross sections (macroscopic):

$$D = 1.263 \text{ cm}$$

$$\sigma_a = 0.0819 \text{ cm}^{-1}$$

$$v\sigma_f = 0.214 \text{ cm}^{-1}$$

We can further re-arrange the equation

$$\nabla^2\phi + \frac{v\sigma_f\phi/k - \sigma_a\phi}{D} = 0$$

$$B^2 = \left(\frac{v\sigma_f/k - \sigma_a}{D}\right)$$

$$\nabla^2\phi + B^2\phi = 0$$

We recognize this is indeed the Helmholtz equation, which is a partial differential equation.

However, for a spherically symmetric mass, we can consider only the term in r, which makes the problem an ordinary differential equation.

$$\nabla^2\phi \equiv \frac{1}{r^2}\frac{\partial}{\partial r}\left(r^2\frac{\partial\phi}{\partial r}\right) \quad \text{for spherical symmetry}$$

Therefore, we have, after expanding terms

$$\frac{d^2\phi}{dr^2} + \frac{2}{r}\frac{d\phi}{dr} + B^2\phi = 0$$

At this point, we can either consider a regular singular point and apply the method of Frobenius in the deleted neighborhood about the origin, or we can consider a substitution method to convert the form of the equation to something more readily solvable. Using a substitution method, we can assume

$$\phi = \frac{w}{r}$$

Then we have

$$\frac{d\phi}{dr} = \frac{w'}{r} - \frac{w}{r^2} \quad \text{and} \quad \frac{d^2\phi}{dr^2} = \frac{w''}{r} - \frac{2w'}{r^2} + \frac{2w}{r^3}$$

Substituting in we obtain

$$\frac{w''}{r} - \frac{2w'}{r^2} + \frac{2w}{r^3} + \frac{2}{r}\left(\frac{w'}{r} - \frac{w}{r^2}\right) + B^2 \frac{w}{r} = 0$$

It can be seen that this reduces to: $w'' + B^2 w = 0$ where

$$\lambda^2 + B^2 = 0 \quad \lambda^2 = -B^2$$

$$\lambda = \pm iB$$

$$w(r) = \tilde{A}_1 \cos(Br) + \tilde{A}_2 \sin(Br)$$

$$But \quad \phi(r) = \frac{w}{r} \rightarrow \left[\phi(r) = A_1 \frac{\cos(Br)}{r} + A_2 \frac{\sin(Br)}{r}\right]$$

We must now apply boundary conditions to our plutonium ball. The first one is that we assume the neutron flux is finite at the center, and the second condition is that the flux can be "extrapolated" to a zero value at some point off of the surface of the sphere.

Since the plutonium ball is assumed to be placed in a vacuum, no neutrons return by reflection once they exit the sphere. In this case, Diffusion theory predicts the "extrapolation length" is a distance of 2D beyond the physical radius, or $\tilde{R} = R_0 + 2D$. As previously noted, this treatment is applied in diffusion theory to obtain the "correct" solution on the actual radial boundary of the sphere for a "vacuum" or "zero return current" boundary condition.

For the first boundary condition

$$\lim_{r \to 0} \phi(r) \equiv finite$$

Then

$$\lim_{r \to 0} \frac{A_1 \cos(Br)}{r} + \frac{A_2 \sin(Br)}{r}$$

This implies $A_1 \to 0$; Also, by L'Hospital Rule

$$\lim_{r \to 0} \frac{\sin(Br)}{r} \Rightarrow \lim_{r \to 0} \frac{B \cos(Br)}{1} \quad or \quad \Rightarrow B$$

So we have
$$\phi(r) = \frac{A_2 \sin(Br)}{r}$$

From the second boundary condition

$$\phi(\tilde{R} = R_0 + 2D) = 0 = \frac{A_2 \sin(B\tilde{R})}{\tilde{R}}$$

$$\therefore B \equiv B_n = \frac{n\pi}{\tilde{R}} = \frac{n\pi}{(R_o + 2D)}$$

To determine the critical size, from reactor physics principles, beyond the scope of the discussion here, we need only to consider the *fundamental mode* solution.

Note: The consideration of only the fundamental (n=1) mode to determine a basis for the critical eigenvalue, critical size, and critical mass is permissible because of the rapid time decay of all higher order eigenvalues in time dependent diffusion of prompt fission neutrons in an exactly critical state for prompt fission; for additional details, this is best covered in a formal reactor physics course.

Therefore, considering $n = 1$ is the "Fundamental mode" for neutrons in a critical system, we see there is a relationship between the geometric term and the value of B^2 as defined from the nuclear parameters for the plutonium:

$$B_1^2 = \frac{\pi^2}{(R_o + 2D)^2} = \frac{\left(\nu\sigma_f - \sigma_a\right)}{D} = 0.10459$$

Solving the equation for R_o yields 7.19 cm for this plutonium fuel.

NUCLEAR APPLICATION: Critical Bare Cylinder

We will now apply neutron diffusion theory to solve for the critical radius of a bare cylinder of uranium-235 if it is fashioned into a square cylinder, where twice the radius is equal to the height, $(2R) = (H)$.

$$D = 1.3175 \text{ cm}$$

$$\sigma_a = 0.0722 \text{ cm}^{-1}, \quad \nu\sigma_f = 0.1687 \text{ cm}^{-1}$$

$$\rho = 18.0 \text{ g/cc}$$

And the steady state diffusion equation is again

$$\nabla^2\phi + \frac{\nu\sigma_f\phi/k - \sigma_a\phi}{D} = 0$$

$$B^2 = \left(\frac{\nu\sigma_f/k - \sigma_a}{D}\right)$$

$$\nabla^2\phi + B^2\phi = 0$$

Moreover, we consider (r, z) coordinates with a cylinder with the origin placed at the center. This is shown in the figure below.

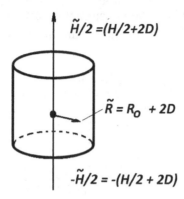

Figure depicting cylindrical geometry

$$\nabla^2 \phi \equiv \frac{1}{r} \frac{\partial}{\partial r}\left(r \frac{\partial \phi}{\partial r}\right) + \frac{\partial^2 \phi}{\partial z^2} \quad \text{for cylindrical geometry}$$

So that the time independent diffusion equation is

$$\frac{1}{r} \frac{\partial}{\partial r}\left(r \frac{\partial \phi}{\partial r}\right) + \frac{\partial^2 \phi}{\partial z^2} + B^2 \phi = 0$$

which expands to become

$$\frac{1}{r}\left(\frac{\partial \phi}{\partial r} + r \frac{\partial^2 \phi}{\partial r^2}\right) + \frac{\partial^2 \phi}{\partial z^2} + B^2 \phi = 0$$

Boundary Conditions

The following boundary conditions apply, and amount to employing a vacuum boundary condition, extrapolating the neutron flux to zero at each outer surface; in addition, the net current should be zero at the geometric center (which is equivalent to a maximum flux condition at the center):

$$\phi(\tilde{R}, z) = 0$$

$$\phi(r, \frac{\tilde{H}}{2}) = 0$$

$$\phi(r, -\frac{\tilde{H}}{2}) = 0$$

$$J\mid_{(0,0)} \hat{n} = -D\nabla\phi\mid_{(0,0)} \cdot \hat{n} = 0$$

According to Separation of Variables, we then assume the following:

$$\phi(r, z) = U(r)V(z)$$

Which leads to

$$\left[\frac{1}{r}(U'V + rU''V) + V''U + B^2 UV = 0\right]\frac{1}{UV}$$

$$\frac{1}{r}\left(\frac{U'}{U} + r\frac{U''}{U}\right) + \frac{V''}{V} + B^2 = 0$$

$$\left(\frac{U''}{U} + \frac{1}{r}\frac{U'}{U}\right) + B^2 + \frac{V''}{V} = 0$$

We then separate each component term and introduce a first separation constant

$$\frac{U''}{U} + \frac{1}{r}\frac{U'}{U} + B^2 = -\frac{V''}{V}$$

To obtain $\quad -\dfrac{V''}{V} = \alpha^2 \quad$ or $\quad \dfrac{V''}{V} = -\alpha^2$

Solving this equation for homogeneous boundaries on the top and bottom of the cylinder, we note that the only plausible solution is

$$V(z) = A_1 \cos(\alpha z) + A_2 \sin(\alpha z)$$

Next we consider the r dimension term and introduce a second separation constant:

$$\frac{U''}{U} + \frac{1}{r}\frac{U'}{U} + B_o^2 - \alpha^2 = 0$$

And let

$$\frac{U''}{U} + \frac{1}{r}\frac{U'}{U} = -\gamma^2$$

Then we have

$$U'' + \frac{1}{r}U' = -\gamma^2 U$$

Multiplying by r^2 we obtain

$$r^2 U'' + rU' = -\gamma^2 r^2 U \quad \text{or} \quad \left[r^2 U'' + rU' + \gamma^2 r^2 U = 0 \right]$$

At this point, we recall the form of an ordinary Bessel's differential equation

$$x^2 y'' + xy' + \left(\lambda^2 x^2 - n^2 \right) y = 0$$

We recall that the general solution to this equation, after applying the Method of Frobenius to a regular singular point in the deleted neighborhood of $x \to 0$ has a series solution given by

$$J_n(x) = \sum_{n=0}^{\infty} \frac{(-1)^r \left(x/2 \right)^{n+2r}}{r! \, \Gamma(n+r+1)}$$

Where the Gamma Function is defined as

$$\Gamma(n+1) = \int_0^\infty x^n e^{-x} dx \quad \text{with} \quad \Gamma(n+1) = n\Gamma(n), \quad \text{or} \quad \Gamma(n+1) = n!$$

The general solution to the ordinary Bessel's equation, after finding a first and a second solution, is therefore

$$\left[y = C_1 J_n(\lambda x) + C_2 Y_n(\lambda x) \right]$$

Returning to the neutron diffusion problem, we note that

$$r^2 U'' + rU' + \gamma^2 r^2 U = 0$$

is an ordinary Bessel's equation of order zero, and the solution is

$$U(r) = C_1 J_o(\gamma r) + C_2 Y_o(\gamma r)$$

Recall that the relationship of separation constants in the PDE yield:

$$\left(-\gamma^2 + B_o^2 - \alpha^2 \right) = 0$$

Then we can solve this equation as follows:

$$\left[B_o^2 = \left(\alpha^2 + \gamma^2\right)\right]$$

Assembling the equations according to separation of variables, we have:

$$\left[\phi(r,z) = \left(C_1 J_o(\gamma r) + C_2 Y_o(\gamma r)\right)\left(A_1 \cos(\alpha z) + A_2 \sin(\alpha z)\right)\right]$$

It is apparent with boundary condition (iv) that

$$C_2 \rightarrow 0 \quad \text{and} \quad A_2 \rightarrow 0$$

since $Y_o(0)$ is unbounded at the origin, and a sine is inconsistent with the origin and homogeneous boundaries along the z-axis. Then we have

$$\phi(r,z) = C J_o(\gamma r)\cos(\alpha z)$$

Applying the radial extrapolated condition (i):

$$\phi\left(\tilde{R},z\right) = 0 = C J_o\left(\gamma \tilde{R}\right)\cos(\alpha z)$$

The only possible solution is to hold that this solution applies to zeros (roots) of the J_o Bessel, which defines the radial eigenvalues:

$$\therefore \gamma \rightarrow \gamma_n = \left(\frac{\xi_n}{\tilde{R}}\right) \text{ where } \xi_n \rightarrow \text{ zeros (roots) of the } J_o \text{ Bessel}$$

Also, with $\phi\left(r, \pm \tilde{H}/2\right) = 0$

we obtain $\therefore \alpha \rightarrow \alpha_m = \left(\frac{m\pi}{\tilde{H}}\right)$ (m odd)

And the solution becomes

$$\phi(r,z) = \sum_{m=1odd}^{\infty} \sum_{n=1}^{\infty} C_{m,n} J_o\left(\frac{\xi_n r}{\tilde{R}}\right) \cos\left(\frac{m\pi z}{\tilde{H}}\right)$$

At this point, we will again specify that for criticality studies (as justified in reactor physics), we can consider only the fundamental mode for steady state criticality, where 2.405 is the first root of the J_o Bessel.

$$\therefore m = n = 1$$

$$\xi_1 = 2.405$$

Assuming the fundamental mode represents the only solution, we obtain

$$\phi(r,z) = C J_o\left(\frac{2.405r}{\tilde{R}}\right)\cos\left(\frac{\pi z}{\tilde{H}}\right)$$

Where the one remaining constant C is determined by the maximum flux in the reactor. It is important to recall that the following relationship still holds:

$$B_o^2 = \alpha^2 + \gamma^2$$

$$\gamma_1 = \frac{2.405}{\tilde{R}}, \quad \alpha_1 = \frac{\pi}{\tilde{H}}$$

Then *critical condition* becomes, for fundamental mode...

$$B_o^2 = \left(\frac{\nu\sigma_f - \sigma_a}{D}\right) = \left(\frac{2.405}{\tilde{R}}\right)^2 + \left(\frac{\pi}{\tilde{H}}\right)^2 = \alpha_1^2 + \gamma_1^2$$

Now, if this is a square cylinder $2R = H$, and $\tilde{H} = H + 4D$ we have

$$\left(\frac{\nu\sigma_f - \sigma_a}{D}\right) = \left(\frac{2.405}{R_o + 2D}\right)^2 + \left(\frac{\pi}{2R_o + 4D}\right)^2$$

Solving this equation for the critical Radius, with the initial values defined for this problem, we get

$$R \rightarrow 7.978\,cm \text{ and } H \rightarrow 15.95\,cm$$

This is the critical size for the square cylinder based on diffusion theory using a fundamental mode steady state solution derived from separation of variables. Again, the theory to develop extrapolated boundary conditions and the fundamental mode requirement for steady state conditions come directly from reactor physics, and are beyond the scope of the discussion here. It is recommended that the reader research a reactor physics text for further elaboration of this topic.

NUCLEAR APPLICATION: Multi-region Criticality Problems

In the previous discussion, we only dealt with a neutron diffusion problem for a single region of fuel surrounded by a vacuum, and thereby applied simple "extrapolated flux" boundary conditions. Problems containing multiple regions of different types of materials require the differential equation to be applied in all regions, and conservation conditions to be applied at the interfaces. These conservation conditions are mathematically known as (i) conservation of solution, and (ii) conservation of gradient. In other words, at an interface, the solutions rendered for a differential equation in each region must be held consistent in the limit as one approaches the boundary. This applies to both the solution and (proportional to) the gradient of the solution at the interface.

For example, in the conduction heat between two solid materials at an interface, we must have a continuity of temperature, and a continuity of heat flux (which is related for each solid by the product of the thermal conductivity and negative temperature gradient in each region, applied in the limit as one approaches the boundary in either material.

In neutron diffusion problems, we must have a continuity of flux (solution) and a continuity of current (neutron flow) at each material interface. If we were to expand the plutonium ball spherical problem into one where we add a second region that is a neutron reflector outside of the plutonium ball considered, we must then consider *two* neutron diffusion equations: the neutron multiplication eigenvalue formulation already considered, and a new "reflector" region (that does not contain any fissile material) outside the plutonium ball.

Assuming we will have the ball reflected by packing peanuts (as it might be in a shipping container), in region 1 (plutonium ball) we have:

$$\frac{d^2\phi_1}{dr^2} + \frac{2}{r}\frac{d\phi_1}{dr} + B^2\phi_1 = 0$$

where now
$$B^2 = \left(\frac{v\sigma_{f1}/k - \sigma_{a1}}{D_1}\right)$$

and in region 2, we have, for example, for the packing peanuts with

$$\frac{d^2\phi_2}{dr^2} + \frac{2}{r}\frac{d\phi_2}{dr} - \frac{\phi_2}{L^2} = 0 \quad \text{where} \quad L^2 = \left(\frac{D_2}{\sigma_{a2}}\right)$$

This is a two-region problem, and a finite reflector region will then have a vacuum "extrapolated" boundary outside of it. Therefore, for neutrons at the interface between regions 1 and 2, the following two equations continuity of (i) flux and (ii) current apply at the interface radius $\vec{r}_s$:

$$\text{(i)}\ \phi_1(\vec{r}_s) = \phi_2(\vec{r}_s); \quad \text{(ii)}\ -D_1\nabla\phi_A\big|\cdot\hat{r}_s = -D_2\nabla\phi_B\big|\cdot\hat{r}_s$$

These conditions compliment the fact that the flux must be finite at the origin, and a vacuum boundary condition. Should the reflector region (e.g. packing peanuts) be infinite, then the flux must be real, bounded, and decreasing as the radius tends toward infinity.

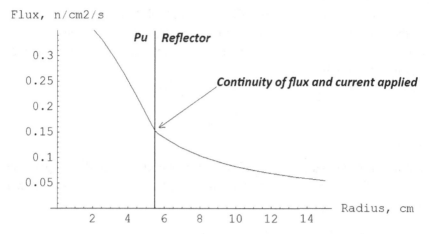

An example of the solution for a two region criticality problem in spherical geometry. All boundary conditions combine to yield a critical condition which nets the value of the minimum (fundamental mode) critical radius at the interface. Following this, the flux was plotted to depict the solution. Continuity of flux matches the fluxes at the interface, but the slopes differ in the two regions based on the diffusion constant of the plutonium and the reflector that differ.

NUCLEAR APPLICATION: *Reflected Cylinder*

Consider a radially reflected, radially symmetric, cylindrical reactor depicted below:

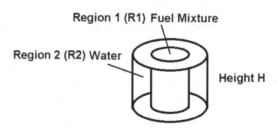

To solve for the criticality condition for this reflected cylinder, we can solve the steady state 1-speed diffusion equation to analytically solve for the critical size and mass (outer radius R2 and height H). Boundary conditions are Vacuum on top (+H/2), bottom (-H/2), and outer radius (R2) of the reactor vessel. Other design constraints include a design keff=1.3 with a fundamental reactor mode to account for initial fuel loading, extrapolated dimensions:

$$\tilde{H} = \frac{2}{3}\tilde{R}_2$$

Other constraints are to use fuel region transport corrected extrapolation distances for *entire* top and bottom surfaces, as well as the constraint that $R_1 = 0.6R_2$:

$$\text{In Region 1:} \quad \frac{1}{r}\frac{\partial}{\partial r}\left(r\frac{\partial \phi_I}{\partial r}\right) + \frac{\partial^2 \phi_I}{\partial z^2} + B^2 \phi_I = 0$$

$$\text{In Region 2:} \quad \frac{1}{r}\frac{\partial}{\partial r}\left(r\frac{\partial \phi_{II}}{\partial r}\right) + \frac{\partial^2 \phi_{II}}{\partial z^2} - \frac{\phi_{II}}{L^2} = 0$$

where

$$L^2 = \frac{D_{II}}{\Sigma_{all}} \qquad B^2 = \frac{\left(\nu\Sigma_{fI}/k_{eff}\right) - \Sigma_{aI}}{D_I}$$

Where Region 1 will result in an ordinary Bessel and a Cosine, and Region 2 will result in a modified Bessel and a Cosine.

In formulating the solution, one must apply conditions that the net current is zero at the problem center (due to symmetry), and extrapolated (vacuum) boundaries on external surfaces. With an application of continuity of flux and current applied at the fuel/reflector interface, this results in a *criticality condition* where:

$$\gamma^2 = B^2 - \frac{\pi^2}{\tilde{H}^2} \qquad \mu^2 = \frac{\pi^2}{\tilde{H}^2} + \frac{1}{L^2}$$

$$D_I \gamma J_1(\gamma R_1) + D_{II} \left(\frac{J_0(\gamma R_1)}{I_0(\mu R_1) - \frac{I_0(\mu \tilde{R}_2)}{K_0(\mu \tilde{R}_2)} K_0(\mu R_1)} \right) \left(\mu I_1(\mu R_1) + \frac{I_0(\mu \tilde{R}_2)}{K_0(\mu \tilde{R}_2)} \mu K_1(\mu R_1) \right) = 0$$

The fluxes in Regions 1 (fuel) and 2 (reflector) are, respectively:

$$\phi_I(r,z) = J_0(\gamma r) \cos(\frac{\pi z}{\tilde{H}})$$

$$\phi_{II}(r,z) = \left(\frac{J_0(\gamma R_1)}{I_0(\mu R_1) - \frac{I_0(\mu \tilde{R}_2)}{K_0(\mu \tilde{R}_2)} K_0(\mu R_1)} \right) \left(I_0(\mu r) - \frac{I_0(\mu \tilde{R}_2)}{K_0(\mu \tilde{R}_2)} K_0(\mu r) \right) \cos(\frac{\pi z}{\tilde{H}})$$

Considering this reflected reactor with 2-region fast spectrum neutron cross sections for Region 1, with a design $k_{eff} = 1.3$, and 19.0 g/cc uranium metal fuel at 20 w% U-235 enrichment, reflected by Region 2, water (1.0 g/cc), using the following absorption, fission production, and transport cross sections and diffusion constant data in each regions, respectively:

$$\Sigma_{a\,I} = 0.02444 \text{ cm}^{-1}, \qquad \nu\Sigma_{f\,I} = 0.04829 \text{ cm}^{-1},$$

$$\Sigma_{tr\,I} = 0.30059 \text{ cm}^{-1}, \qquad D_I = 1.10893 \text{ cm}$$

$$\Sigma_{a\,II} = 0.00011 \text{ cm}^{-1}, \qquad \nu\Sigma_{f\,II} = 0.0000 \text{ cm}^{-1},$$

$$\Sigma_{tr\,II} = 0.23034 \text{ cm}^{-1}, \qquad D_{II} = 1.44715 \text{ cm}$$

Root solving the critical condition equation yields an extrapolated radius of $\tilde{R}_2 = 54.14$ cm; this leads to a physical radius of $R_2 = 51.78$ cm, and a height $H = 31.37$ cm. A normalized plot of the flux is shown in the figure below.

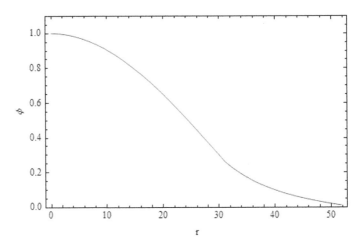

Normalized plot of the center flux for a reflected reactor system at z=0; the reflector region (Region 2) begins at r=31.067 cm.

Chapter 13

Numerical Solutions of Partial Differential Equations

Due to their complexity, which is evident for seemingly simple problems, partial differential equations are often best solved numerically. This chapter presents the fundamental knowledge needed in order to derive the systems of equations needed for numerical finite difference solutions of PDEs. These systems of equations can then be directly solved via linear algebra methods discussed previously.

Two Approaches to Forming Finite Difference Equations

The two approaches to deriving finite difference equations are known as the Control Volume Method (CVM) and the Differential Equations Method (Schmidt's Method). Both can be used to render finite difference equations in a fully consistent manner. The CVM approach uses a direct voxelized integration of the PDE, while Schmidt's Method is a "nodal" approach using Taylor's Series and is highly flexible. Both methods can be associated with a surrounding control volume, explicitly with the Control Volume Method, and implicitly with Schmidt's Method.

CVM Summary

The procedure for the Control Volume (CV) Method is as follows:

- subdivide entire problems into small CVs
- locate "nodes" at volume center to represent solution in CV
- integrate conservation equation over CV

Consider a Cartesian coordinate system with the familiar compass notation for a control volume with a central general "o" node placed in the center of the node. The function value at the node is assumed to apply throughout the control volume, which is further assumed to approximate a differential volume.

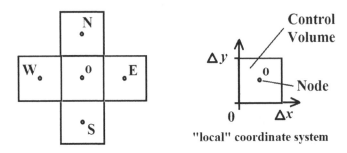

Consider the one energy neutron diffusion equation:

$$\frac{\partial^2 \phi}{\partial x^2} + \frac{\partial^2 \phi}{\partial y^2} - \frac{\Sigma_a \phi}{D} + \frac{Q_0}{D} = 0$$

Using the Control Volume Method (CVM) of deriving the finite difference formulations, we integrate the entire PDE over the local coordinate system

$$\int_0^{\Delta y} \int_0^{\Delta x} \left(\frac{\partial^2 \phi}{\partial x^2} + \frac{\partial^2 \phi}{\partial y^2} - \frac{\Sigma_a \phi}{D} + \frac{Q_0}{D} \right) dx\, dy = 0$$

Then we expand derivatives

$$\int_0^{\Delta y} \int_0^{\Delta x} \left(\frac{\partial}{\partial x} \left(\frac{\partial \phi}{\partial x} \right) + \frac{\partial}{\partial y} \left(\frac{\partial \phi}{\partial y} \right) - \frac{\Sigma_a \phi}{D} + \frac{Q_0}{D} \right) dx\, dy = 0$$

Following this, we integrate each term:

$$\int_0^{\Delta y}\int_0^{\Delta x} \frac{\partial}{\partial x}\left(\frac{\partial \phi}{\partial x}\right) dx\, dy + \int_0^{\Delta x}\int_0^{\Delta y} \frac{\partial}{\partial y}\left(\frac{\partial \phi}{\partial y}\right) dy\, dx$$

$$+ \int_0^{\Delta x}\int_0^{\Delta y} \frac{-\Sigma_a \phi}{D} dx\, dy + \int_0^{\Delta x}\int_0^{\Delta y} \frac{Q_o}{D} dx\, dy = 0$$

This then simplifies to

$$\int_0^{\Delta y} \left.\frac{\partial \phi}{\partial x}\right]_0^{\Delta x} dy + \int_0^{\Delta x} \left.\frac{\partial \phi}{\partial y}\right]_0^{\Delta y} dx - \frac{\Sigma_a \phi_o}{D}\Delta x\, \Delta y + \frac{Q_o}{D}\Delta x\, \Delta y = 0$$

We must maintain positive axis sense... $+ x, + y$ relative to all derivatives.

$$\left(\left.\frac{\partial \phi}{\partial x}\right|_{\Delta x} - \left.\frac{\partial \phi}{\partial x}\right|_o\right)\Delta y + \left(\left.\frac{\partial \phi}{\partial y}\right|_{\Delta y} - \left.\frac{\partial \phi}{\partial y}\right|_o\right)\Delta x - \frac{\Sigma_a \phi_o \Delta x \Delta y}{D} + \frac{Q_o}{D}\Delta x\, \Delta y = 0$$

or

$$\left(\frac{\phi_E - \phi_o}{\Delta x} - \frac{\phi_o - \phi_w}{\Delta x}\right)\Delta y + \left(\frac{\phi_N - \phi_o}{\Delta y} + \frac{\phi_o - \phi_s}{\Delta y}\right)\Delta x - \frac{\Sigma_a \phi_o \Delta x \Delta y}{D} + \frac{Q_o}{D}\Delta x\, \Delta y = 0$$

Then, we can multiply by

$$\frac{1}{\Delta x \Delta y}$$

and combine terms with some simplification we obtain the finite difference formulation for the PDE:

$$\left[\frac{\phi_E - 2\phi_o + \phi_w}{\Delta x^2} + \frac{\phi_N - 2\phi_o + \phi_s}{\Delta y^2} - \frac{\Sigma_a \phi_o}{D} + \frac{Q_o}{D} = 0\right]$$

This becomes the general node finite difference equation for the control volume with cell centered nodes.

The Differential Equation Method (Schmidt's Method)

This alternate method is more powerful for deriving finite difference formulations. It is patterned after Dr Frank Schmidt's method for deriving any finite difference formulation for any coordinate system, and is based fundamentally on the application of Taylor's series computed over a grid of nodes separated by either uniform or non-uniform intervals. The differencing equations rendered using Schmidt's method are identical to those from the CVM, however, they also provide the truncation error of the differencing formula approximation of the derivatives. A "control volume" can be therefore associated with a differential equation derived at a particular point or "node", where the "nodal differencing equation" approximates that equation. A set of nodes with associated "control volumes" surrounding them are depicted in the figure below:

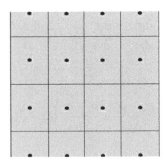

Nodes where differencing equations are to be derived using the Taylor's Series by applying Schmidt's method. Nodes are shown equally spaced, although they are not required to be; unequally spaced nodes are easily handled using Schmidt's method.

Schmidt's Method Summary

The procedure for the Differential Equation (Schmidt's) Method is as follows:

- Subdivide entire problem into nodes with implicit control volumes surrounding nodes; write the equations for surrounding nodes along orthogonal directions using the Taylor's Series
- Introduce weighting coefficients in each direction for a given step size

- Derive the equations for and solve for weighting coefficients based on the PDE being solved; sum equations to yield differencing equations and truncation error formulas

Recall the Taylor's Series is given as

$$f(x) \approx f(x_o) + f'(x_o)(x - x_o) + \frac{f''(x_o)}{2!}(x - x_o)^2 + \ldots$$

Apply Taylor's Series at each node surrounding "o" node and assemble terms of differential equation using weighting coefficients.

Begin with the steady state diffusion partial equation

$$\frac{\partial^2 \phi}{\partial x^2} + \frac{\partial^2 \phi}{\partial y^2} - \frac{\Sigma_a \phi}{D} + \frac{Q_o}{D} = 0$$

Then at the "o" node

$$\left. \frac{\partial^2 \phi}{\partial x^2} \right|_o + \left. \frac{\partial^2 \phi}{\partial y^2} \right|_o + \frac{-\Sigma_{ao}\phi_o}{D_o} + \frac{Q_o}{D_o} = 0$$

Now assemble Taylor's Series Expansions with weighting coefficients. Since the Cartesian grid pattern is orthogonal, we initially consider only x derivative terms in with weighting coefficients A and B to be determined.

$$A \left[\phi_E = \phi_o + \left. \frac{\partial \phi}{\partial x} \right|_o \Delta x + \left. \frac{\partial^2 \phi}{\partial x^2} \right|_o \frac{\Delta x^2}{2!} + \left. \frac{\partial^3 \phi}{\partial x^3} \right|_o \frac{\Delta x^3}{3!} + \left. \frac{\partial^4 \phi}{\partial x^4} \right|_o \frac{\Delta x^4}{4!} + \ldots \right]$$

$$B \left[\phi_w = \phi_o + \left. \frac{\partial \phi}{\partial x} \right|_o (-\Delta x) + \left. \frac{\partial^2 \phi}{\partial x^2} \right|_o \frac{(-\Delta x)^2}{2!} + \left. \frac{\partial^3 \phi}{\partial x^3} \right|_o \frac{(-\Delta x)^3}{3!} + \left. \frac{\partial^4 \phi}{\partial x^4} \right|_o \frac{(-\Delta x)^4}{4!} + \ldots \right]$$

Recall from the PDE there is only a single term with derivatives in x, and we must match the constraints on the coefficient equations using the original PDE,

$$\left[\sum Coeffs \left. \frac{\partial^2 \phi}{\partial r^2} \right|_o = 1 \right] \qquad \left[\sum Coeffs \left. \frac{\partial \phi}{\partial r} \right|_o = 0 \right]$$

Which leads directly to

$$\text{(i)} \quad A\frac{\Delta x^2}{2!} + B\frac{\Delta x^2}{2!} = 1 \qquad \text{and} \qquad \text{(ii)} \quad A\Delta x + B(-\Delta x) = 0$$

Therefore the procedure yields two equations, two unknowns. From these two equations we have

$$A\Delta x = B\Delta x \quad \text{or} \quad (A = B)$$

$$A\frac{\Delta x^2}{2!} + A\frac{\Delta x^2}{2!} = 1 \quad \text{or} \quad 2A\frac{\Delta x^2}{2!} = 1$$

Therefore, we obtain $A = \dfrac{1}{\Delta x^2} \quad B = \dfrac{1}{\Delta x^2}$ which results in

$$\frac{\phi_E}{\Delta x^2} = \frac{\phi_o}{\Delta x^2} + \frac{1}{\Delta x}\frac{\partial\phi}{\partial x}\bigg|_o + \frac{1}{2!}\frac{\partial^2\phi}{\partial x^2}\bigg|_o + \frac{1}{3!}\frac{\partial^3\phi}{\partial x^3}\bigg|_o \Delta x + \frac{1}{4!}\frac{\partial^4\phi}{\partial x^4}\bigg|_o \Delta x^2 + \dots$$

$$\frac{\phi_w}{\Delta x^2} = \frac{\phi_o}{\Delta x^2} + \frac{-1}{\Delta x}\frac{\partial\phi}{\partial x}\bigg|_o + \frac{1}{2!}\frac{\partial^2\phi}{\partial x^2}\bigg|_o + \frac{-1}{3!}\frac{\partial^3\phi}{\partial x^3}\bigg|_o \Delta x + \frac{1}{4!}\frac{\partial^4\phi}{\partial x^4}\bigg|_o \Delta x^2 + \dots$$

Summing these two equations gives

$$\frac{\phi_E + \phi_w}{\Delta x^2} = \frac{2\phi_o}{\Delta x^2} + 0 + \frac{\partial^2\phi}{\partial x^2}\bigg|_o + \frac{\Delta x^2}{12}\frac{\partial^4\phi}{\partial x^4}\bigg|_o + \dots$$

Solve for $\dfrac{\partial^2\phi}{\partial x^2}\bigg|_o$ to yield $\left[\dfrac{\partial^2\phi}{\partial x^2} = \dfrac{\phi_E - 2\phi_o + \phi_w}{\Delta x^2} - \dfrac{\Delta x^2}{12}\dfrac{\partial^4\phi}{\partial x^4}\bigg|_o + \dots\right]$

Similarly, for the *y*-direction

$$\left[\frac{\partial^2\phi}{\partial y^2} = \frac{\phi_N - 2\phi_o + \phi_s}{\Delta y^2} - \frac{\Delta y^2}{12}\frac{\partial^4\phi}{\partial y^4}\bigg|_o + \dots\right]$$

Then the algebraic approximation to the PDE is, at each general node:

$$\left[\frac{\phi_E - 2\phi_o + \phi_w}{\Delta x^2} + \frac{\phi_N - 2\phi_o + \phi_s}{\Delta y^2} - \frac{\Sigma_{ao}\phi_o}{D_o} + \frac{Q_o}{D_o} - \left(\frac{\Delta x^2}{12} \frac{\partial^4 \phi}{\partial x^4} \Big| + \frac{\Delta y^2}{12} \frac{\partial^4 \phi}{\partial y^4} \Big| + ... \right) = 0 \right]$$

Note that Schmidt's method yields the identical algebraic finite difference equations as in the CVM approach, but now we also have the *truncation error* of the algebraic approximation. We can apply this as needed to estimate gridsizes to yield appropriate accuracy goals.

Also, Schmidt's method can be applied at any collection of points, and can be used to combine non-orthogonal gridpoints using the multivariate form of the Taylor's series. Overall, this makes Schmidt's method very powerful—any finite difference formulation can be solved for in any application using this approach. Note that we can also derive this in other coordinate systems by handling the $\left(\nabla^2 \phi \right)$ term to yield the appropriate differencing formulas in the diffusion equation.

$$D\nabla^2 \phi - \Sigma_a \phi + \frac{\nu \Sigma_f \phi}{k} + Q_o = 0$$

Schmidt's Method in 2-D Cylindrical Coordinates

In this case, the two dimensional neutron diffusion equation is

$$\frac{\partial^2 \phi}{\partial r^2} + \frac{1}{r} \frac{\partial \phi}{\partial r} + \frac{\partial^2 \phi}{\partial (r\theta)^2} - \frac{\Sigma_a \phi}{D} + \frac{Q_o}{D} = 0$$

As before, we can assemble Taylor's Series Expansions with weighting coefficients. The cylindrical grid pattern is an orthogonal curvilinear coordinate system, and we consider derivative terms in with weighting coefficients *A, B, C,* and *E* to be determined.

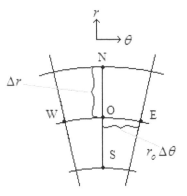

Geometry for r–z grid differencing

$$A \quad \left[\phi_E \;=\; \phi_o + \frac{1}{r_o}\frac{\partial \phi}{\partial \theta}\bigg|_o r_o\Delta\theta + \frac{1}{r_o^2}\frac{\partial^2 \phi}{\partial \theta^2}\bigg|_o \frac{(r_o\Delta\theta)^2}{2!} + \ldots \right]$$

$$B \quad \left[\phi_w \;=\; \phi_o + \frac{1}{r_o}\frac{\partial \phi}{\partial \theta}\bigg|_o (-r_o\Delta\theta) + \frac{1}{r_o^2}\frac{\partial^2 \phi}{\partial \theta^2}\bigg|_o \frac{(-r_o\Delta\theta)^2}{2!} + \ldots \right]$$

$$C \quad \left[\phi_N \;=\; \phi_o + \frac{\partial \phi}{\partial r}\bigg|_o \Delta r + \frac{\partial^2 \phi}{\partial r^2}\bigg|_o \frac{\Delta r^2}{2!} + \ldots \right]$$

$$E \quad \left[\phi_s \;=\; \phi_o + \frac{\partial \phi}{\partial r}\bigg|_o (-\Delta r) + \frac{\partial^2 \phi}{\partial r^2}\bigg|_o \frac{(-\Delta r)^2}{2!} + \ldots \right]$$

The coefficients multiplying the Taylor's Series terms must be assembled so as to pattern the requirements specified by the original differential equation:

$$\left[\sum Coeffs \; \frac{\partial^2 \phi}{\partial r^2}\bigg|_o = 1 \right], \qquad\qquad \left[\sum Coeffs \; \frac{\partial \phi}{\partial r}\bigg|_o = \frac{1}{r_o} \right]$$

$$\left[\sum Coeffs \; \frac{\partial^2 \phi}{\partial \theta^2}\bigg|_o = \frac{1}{r_o^2} \right], \qquad\qquad \left[\sum Coeffs \; \frac{\partial \phi}{\partial \theta}\bigg|_o = 0 \right]$$

We find that we can solve for each orthogonal variable along each axis direction. In the radial direction, we solve these equations simultaneously to obtain

$$\left[C = \frac{2r_o + \Delta r}{2r_o \Delta r^2} \right] \left[E = \frac{2r_o - \Delta r}{2r_o \Delta r^2} \right]$$

It should be noted that these are different due to the relative area difference in the curvilinear model, to account for the changing radius. With derived coefficients, multiply and sum the equations to yield the differencing formulation in the along the radial orthogonal axis:

$$\frac{2r_o + \Delta r}{2r_o \Delta r^2}\left(\phi_N = \phi_o + \frac{\partial \phi}{\partial r}\bigg|_o \Delta r + \frac{\partial^2 \phi}{\partial r^2}\bigg|_o \frac{\Delta r^2}{2!} + ... \right)$$

$$\frac{2r_o - \Delta r}{2r_o \Delta r^2}\left(\phi_S = \phi_o + \frac{\partial \phi}{\partial r}\bigg|_o (-\Delta r) + \frac{\partial^2 \phi}{\partial r^2}\bigg|_o \left(\frac{(-\Delta r)^2}{2!} \right) + ... \right)$$

Summing, we obtain

$$\phi_N \left(\frac{2r_o + \Delta r}{2r_o \Delta r^2} \right) + \phi_S \left(\frac{2r_o - \Delta r}{2r_o \Delta r^2} \right) = \frac{4r_o \phi_o}{2r_o \Delta r^2} + \frac{4r_o \Delta r^2}{2r_o \Delta r^2 2!} \frac{\partial^2 \phi}{\partial r^2}\bigg|_o$$

Then, for the radial direction, we have

$$\left[\frac{\partial^2 \phi}{\partial r^2}\bigg|_o + \frac{1}{r_o}\frac{\partial \phi}{\partial r}\bigg|_o = \phi_N \left(\frac{2r_o - \Delta r}{2r_o \Delta r^2} \right) + \phi_S \left(\frac{2r_o - \Delta r}{2r_o \Delta r^2} \right) - \frac{2\phi_o}{\Delta r^2} + \varepsilon_r \right]$$

It can be shown by carrying out more terms in the Taylor's Series, the radial truncation error is also affected directly by the radius of the gridpoint:

$$\left[\varepsilon_r = -\left(\frac{\Delta r^2}{3!r_o} \right) \left. \frac{\partial^3 \phi}{\partial r^3} \right|_o + ... \right]$$

In the "θ" direction, it can be shown that

$$\left[A = B = \frac{1}{(r_o \Delta \theta)^2} \right]$$

$$\left[\left. \frac{\partial^2 \phi}{\partial (r\theta)^2} \right|_o = \frac{\phi_E}{(r_o \Delta \theta)^2} + \frac{\phi_W}{(r_o \Delta \theta)^2} - \frac{2\phi_o}{(r_o \Delta \theta)^2} + \varepsilon_\theta \right]$$

The truncation term in the "θ" direction can be shown to be

$$\left[\varepsilon_\theta = -\frac{2(\Delta \theta)^2}{4! r_o^2} \left. \frac{\partial^4 \phi}{\partial \theta^4} \right|_o + ... \right]$$

Assembling terms, the neutron diffusion finite difference equations in 2-D cylindrical coordinates is

$$[\ \phi_N \left(\frac{2r_o + \Delta r}{2r_o \Delta r^2} \right) + \phi_S \left(\frac{2r_o - \Delta r}{2r_o \Delta r^2} \right) + \frac{\phi_E + \phi_W}{(r_o \Delta \theta)^2} - 2\phi_o \left(\frac{1}{\Delta r^2} + \frac{1}{(r_o \Delta \theta)^2} \right)$$

$$-\frac{\Sigma_{ao} \phi_o}{D_o} + \frac{Q_o}{D_o} + \varepsilon = 0 \ ]$$

The truncation error of this formulation is

$$\left[\varepsilon = \varepsilon_r + \varepsilon_\theta = -\left(\frac{\Delta r^2}{3!r_o} \right) \left. \frac{\partial^3 \phi}{\partial r^3} \right|_o - \frac{2(\Delta \theta)^2}{4! r_o^2} \left. \frac{\partial^4 \phi}{\partial \theta^4} \right|_o + ... \right]$$

We note here there will be difficulties at the node located on the origin, where in this case, $r \to 0$. There are a few approaches that can be performed; a simple one is to approximate the equation at the origin with the Cartesian form of the equation, possible due to the symmetric nature of the origin point; again, this should only be used that node.

Numerical Implementation

Starting with the finite difference equations, we can apply these to each of several nodes to create a linear system of equations be solved for the neutron flux (n/cm²/s) at general grid nodes.

$$\left[\frac{\phi_E - 2\phi_o + \phi_w}{\Delta x^2} + \frac{\phi_N - 2\phi_o + \phi_s}{\Delta y^2} - \frac{\Sigma_a \phi_o}{D} + \frac{Q_o}{D} = 0 \right]$$

The geometry to be used is given below

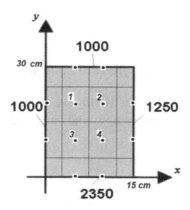

2-D Cartesian Geometry neutron diffusion finite difference problem

For this problem, a uniform material and constant flux boundary conditions (as indicated) are specified. Note also that for this problem, $\Delta x = 5\,\text{cm}$, $\Delta y = 10\,\text{cm}$, $Q_o = 0\ \text{cm}^{-3}\,\text{s}^{-1}$ $D = 1\ \text{cm}$, $\Sigma_a = 1\ \text{cm}^{-1}$.

We can assemble the equations for *each* of the four (4) nodes using matrix notation, where $A\,x = b$. This is done by considering each node, and projecting the compass grid (o, N, S, E, W) on each gridpoint and building the finite difference equations. For the problem given, the finite difference diffusion equaion for the four nodes for this problem results in, using $A\,x = b$ notation:

$$
\begin{pmatrix}
-\dfrac{11}{10} & \dfrac{1}{25} & \dfrac{1}{100} & 0 \\[2mm]
\dfrac{1}{25} & -\dfrac{11}{10} & 0 & \dfrac{1}{100} \\[2mm]
\dfrac{1}{100} & 0 & -\dfrac{11}{10} & \dfrac{1}{25} \\[2mm]
0 & \dfrac{1}{100} & \dfrac{1}{25} & -\dfrac{11}{10}
\end{pmatrix}
\begin{pmatrix}
\phi[1] \\ \phi[2] \\ \phi[3] \\ \phi[4]
\end{pmatrix}
=
\begin{pmatrix}
-50. \\ -60. \\ -63.5 \\ -73.5
\end{pmatrix}
$$

The solution of this system is obtained via standard linear algebra techniques.

Cartesian Albedo Boundary Difference Equations for Diffusion

At a boundary, it is common practive to set a reflection factor, such that at the right hand edge of a problem, as shown in the Figure below for an "East" boundary:

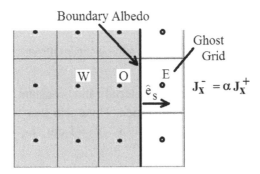

East Boundary "Ghost" gridpoint; the fictitious East gridpoint is used to establish the albedo reflection in a differencing formulation.

Recall we determined that the Laplacian in Cartesian geometry, good for 2nd order truncation, is:

$$\nabla^2 \phi \Big|_o = \frac{\partial^2 \phi}{\partial x^2}\Big|_o + \frac{\partial^2 \phi}{\partial y^2}\Big|_o \approx \frac{\phi_E + \phi_W - 2\phi_o}{\Delta x^2} + \frac{\phi_N + \phi_S - 2\phi_o}{\Delta y^2}$$

The East boundary nodal value for the gridpoint flux ϕ_E must be replaced with a formulation that is determined by the reflection factor, as shown in the figure.

Therefore this can be derived through application of differencing formulas with partial currents (as noted in the previous chapter):

$$J^- = \alpha_R J^+$$

$$\frac{\phi_o}{4} + \frac{D}{2}\frac{d\phi}{dx}\Big|_o = \alpha_R (\frac{\phi_o}{4} - \frac{D}{2}\frac{d\phi}{dx}\Big|_o)$$

$$\frac{\phi_o}{4}(1 - \alpha_R) = \frac{-D}{2}\frac{d\phi}{dx}\Big|_o (1 + \alpha_R)$$

$$\left[-\frac{d\phi}{dx}\Big|_o = \frac{\phi_o(1 - \alpha_R)}{2D(1 + \alpha_R)} \right]$$

A 2nd order truncation Central Difference at "O" node (applying Schmidt's Method) is:

$$\frac{d\phi}{dx}\Big|_o \approx \frac{\phi_E - \phi_W}{2\Delta x}$$

Substituting this into the above formulation and solving for the gridpoint flux ϕ_E:

$$-\frac{(\phi_E - \phi_W)}{2\Delta x} = \frac{\phi_o(1 - \alpha_R)}{2D(1 + \alpha_R)}$$

$$-\phi_E = -\phi_W + \frac{\phi_o \Delta x(1 - \alpha_R)}{D(1 + \alpha_R)}$$

$$\phi_E = \phi_W - \frac{\phi_o \Delta x(1 - \alpha_R)}{D(1 + \alpha_R)}$$

So, for the second derivative with respect to x on the *right* side boundary, the EAST grid is treated as a *"ghost node"*, and we substitute the above albedo relationship into the 2nd order finite difference equation:

$$\frac{d^2\phi}{dx^2}\bigg|_o \approx \frac{\phi_E + \phi_W - 2\phi_o}{\Delta x^2} \rightarrow \frac{\left(\phi_W - \dfrac{\phi_o \Delta x(1 - \alpha_R)}{D(1 + \alpha_R)}\right) + \phi_W - 2\phi_o}{\Delta x^2}$$

$$\rightarrow \frac{d^2\phi}{dx^2}\bigg|_o \approx \frac{2\phi_W - 2\phi_o}{\Delta x^2} - \frac{\phi_o(1 - \alpha_R)}{D\Delta x(1 + \alpha_R)}$$

Similarly, the ghost gridpoint methodology with the partial current relation can be applied in the same manner on the left boundary. The West boundary nodal value for the gridpoint flux ϕ_W must be replaced with a formulation that is determined by the left albedo reflection factor, as indicated below:

$$\frac{d^2\phi}{dx^2}\bigg|_o \approx \frac{\phi_E + \phi_W - 2\phi_o}{\Delta x^2} \rightarrow \frac{\phi_E + \left(\phi_E - \dfrac{\phi_o \Delta x(1 - \alpha_L)}{D(1 + \alpha_L)}\right) - 2\phi_o}{\Delta x^2}$$

$$\rightarrow \frac{d^2\phi}{dx^2}\bigg|_o \approx \frac{2\phi_E - 2\phi_o}{\Delta x^2} - \frac{\phi_o(1 - \alpha_L)}{D\Delta x(1 + \alpha_L)}$$

Moreover, it is readily shown that this methodology can be applied at the bottom (South gridpoint) and top (North gridpoint) as well, respectively:

$$\frac{d^2\phi}{dy^2}\bigg|_o \approx \frac{2\phi_N - 2\phi_o}{\Delta y^2} - \frac{\phi_o(1-\alpha_B)}{D\Delta y(1+\alpha_B)}$$

$$\frac{d^2\phi}{dy^2}\bigg|_o \approx \frac{2\phi_S - 2\phi_o}{\Delta y^2} - \frac{\phi_o(1-\alpha_T)}{D\Delta y(1+\alpha_T)}$$

The substitutions can be made to alter the Laplacian for representing the differential equation for diffusion theory, as required. A similar treatment can be made for any other problems where flow can be computed.

Multigroup Diffusion

Up to this point, we have only alluded to the fact that one can have a multiple energy group treatment, having suggested this when initially presenting the neutron transport equation in the previous chapter.

In the multigroup formulation, neutrons proceed in interactions down the energy grid from Group $g = 1$ (fast) neutrons down to the last group G. As far as energy dependent fluxes are concerned, the group dependent flux is averaged (as are cross sections, etc) so that:

$$\phi_g = \int_{E_g}^{E_{g-1}} dE\,\phi(E)$$

One can also specify a group dependent scalar surce term:

$$Q_g = \int_{E_g}^{E_{g-1}} dE\,Q(E)$$

In that case, the group depenent quantities can be enumerated in the diffusion equation, where we have omitted the theoretical arguments supporting these equations; the multi-group diffusion equations have the form as shown:

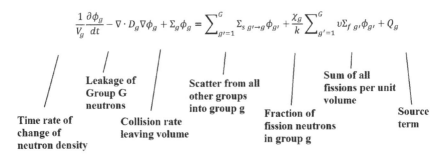

$$\frac{1}{V_g}\frac{\partial \phi_g}{dt} - \nabla \cdot D_g \nabla \phi_g + \Sigma_g \phi_g = \sum_{g'=1}^{G} \Sigma_{s\ g' \rightarrow g} \phi_{g'} + \frac{\chi_g}{k} \sum_{g'=1}^{G} \upsilon \Sigma_{f\ g'} \phi_{g'} + Q_g$$

Time rate of change of neutron density

Leakage of Group G neutrons

Collision rate leaving volume

Scatter from all other groups into group g

Fraction of fission neutrons in group g

Sum of all fissions per unit volume

Source term

Form of the general multigroup neutron diffusion equations

Special attention should be called to the scattering cross section, where now, neutron scattering is an interaction probablility of neutrons that scatter from one group g' to the group of interest g as indicated. Therefore, group constants can be assigned with the differencing formulations as required to enable multi-energy group computation of criticality or fixed source problems.

This concludes the material presented as part of Foundations in Applied Nuclear Engineering Analysis. The Appendix contains problem sets that may be used as effective exercises in learning the many subjects presented in this text.

Appendix

Selected Problem Sets in Applied Nuclear Engineering Analysis

Set #I

1a. One interpretation of particle flux ϕ is that it is the number of particles moving in any direction through a unit area per unit of time. If we consider a point source emanating particles outward without scattering, as we move away from the source the particle flux will drop due to spherical divergence.

Compute the particle flux of photons at 1, 10, and 100 cm away from a tiny seed of a 2.7 microCurie (μCi) source of Co-60, assumed to be placed at the isocenter, or origin, of a large voided space. In doing so, note that

- Co-60 gives off 2 photons per disintegration.
- One Curie (Ci) is equivalent to 3.7E10 disintegrations per second

Give your answers in photons/(cm^2 s)

1b. If you stand 2 meters from the point source, and assuming every photon that strikes you has a 30% probability of being absorbed in your body, estimate how many photons are absorbed.

State all assumptions.

1c. Compute the probability of winning the lottery from either of 5 individually purchased tickets. For each ticket, assume that the buyer must choose any 6 numbers randomly spanning from 1 to 53.

1d. Six cards are drawn at random from a standard deck of 52 cards (4 suits of 2… 10, J, Q, K, A). Using the *Combination formula*, find the probability that *all* of the cards drawn are Hearts.

2. Given: Isotope "X" is used as a radiotherapy source for treating cancer patient tumors. The plot below gives the fraction of isotope "X" remaining versus time in hours.

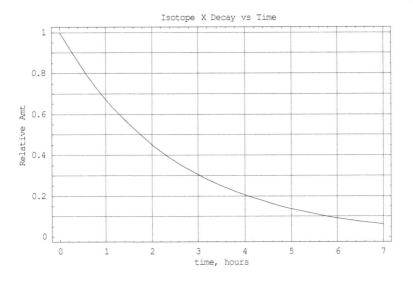

2a. What is the half life $T_{1/2}$ if isotope "X" determined using the plot. State units.

2b. What is the probability (at any time t) that isotope "X" will decay?

2c. Isotope "Y" (a short lived nuclide) and "Z" (a longer lived nuclide) are mixed in a sample, and each has two very different half lives. The plot below gives the fraction of isotopes "Y" and "Z" remaining in the mixed sample versus time in hours.

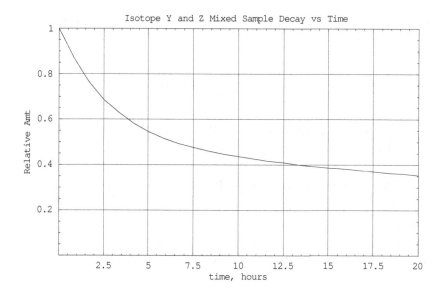

Estimate the half lives $T_{1/2}$ of isotopes "Y" and "Z" using the mixture plot. State units.

3a. Solid Angle.

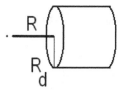

Determine the solid angle of a $R_d = 6$ cm radius cylindrical gamma detector placed at various ranges R away (R values from 2 cm to 100 cm) from a point source of Cs-137. Do this by showing your integration based on the formulation:

$$\Omega = \iint \sin(\theta)\, d\theta\, d\varphi$$

Use plotting software to plot solid angle versus distance. Label plots appropriately.

3b. How much error will occur in the solid angle if you estimate the solid angle using:

$$\Omega \sim \frac{1}{4\pi R^2} \pi R_d^2$$

Plot the error in using this formulation to determine the solid angle versus distance. Label plots appropriately.

4. **Fission Spectra.** Appendix H of the MCNP manual (Ref: LA-UR-03-1987) lists constants for the fission spectra $\chi(E)$ for all known fissile isotopes.

Some nuclides use a Maxwellian Fission Spectrum, where:

$$\chi(E) = C_m E^{1/2} \exp(-E/a_m)$$

Other nuclides use a Watt Fission Spectrum, where:

$$\chi(E) = C_w \exp(-E/a_w) \sinh([b_w E]^{1/2})$$

4a. Regardless of which distribution function is used, the definition of $\chi(E)\, dE$ is:

_____

_____

4b. Consider thermal neutron fission data for the following two fissile nuclides:

For **Pu-241** as defined by a Maxwellian Spectrum:
$a_m = 1.3597$, $C_m = ?$

For **U-235** as defined by a Watt Spectrum:
$a_w = 0.988$, $b_w = 2.249$, $C_w = ?$

Write a code to perform Monte Carlo based numerical integration to estimate the normalization constants (C_m or C_w) for each respective nuclide to permit each distribution to satisfy the required condition that:

$$\int_0^\infty \chi(E)\, dE \equiv 1.$$

(Hint: since *infinity* is not a *useful* upper limit, *justify* a reasonable finite upper limit).

4c. After determining the correct constants, using your TKSolver model for this problem, compute the *average fission neutron energy* for each nuclide, and state how many histories (trials) you used for a relative accuracy of 0.001.

5. Modify your model to perform Monte Carlo based numerical integration to be able to integrate multivariate functions $f(x, y)$ and provide an error estimate.

5a. Use your Monte Carlo multivariate integrator to compute

$$\int_0^1 \int_0^1 \left(x^2 + y^2\right) dx\, dy$$

Report the number of histories so that the estimated sigma is < 0.0001.

5b. Use your TKSolver Monte Carlo multivariate integrator to compute

$$\int_0^{3/2} \int_0^{5/3} \left(\exp(x^2) + \exp(y^2)\right) dx\, dy$$

Note that $\exp(x) = e^x$.

Report the number of histories needed so that the estimated sigma is < 0.0001.

6. Solid Angle Disk Source.

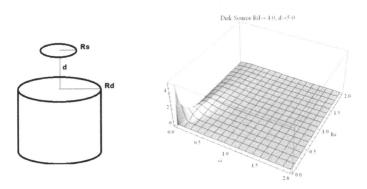

Disk Source Rd→ 4 0, d→5 0

The Solid Angle of a radioactive gamma Disk Source (Disk Source Radius R_s) a distance d away from a circular radiation detector of Radius R_d is computed using the $J_1(x)$ Bessel function according to the formula:

$$\Omega = \lim_{\omega \to \infty} 4\pi \int_0^\omega R_d/R_s \exp(-\frac{\omega d}{R_s}) \frac{J_1(\omega)}{\omega} J_1(\omega R_d/R_s) d\omega$$

$$\Omega_{fraction} = \lim_{\omega \to \infty} \int_0^\omega R_d/R_s \exp(-\frac{\omega d}{R_s}) \frac{J_1(\omega)}{\omega} J_1(\omega R_d/R_s) d\omega$$

Note: A reasonable numerical upper limit for ω is 15, and the Bessel Function can be accurately computed using the truncated power series:

$J_1(x) = (x/2) - (x^3/16) + (x^5/384) - (x^7/18432) + (x^9/1474560)$
$- (x^{11}/176947200) + (x^{13}/29727129600)$

Compute the solid angle by the following methods:

6a. Use TK-Solver with a built in function of syntax: INTEGRAL("domega(w),w",0.0, 'inf)

6b. Write a FORTRAN or C language code or TKSolver procedure function based code using a subroutine for each of the following (see discussions in the course text):

- Midpoint method 'MID' using the midpoint of each sub-interval
- Trapezoidal method 'TRAP' using ½*(LEFT + RIGHT), referring to an average of the function evaluated on the "Left" and "Right" sides of each sub-interval
- Simpsons Rule using the weighted mean of 1/3*(2*MID + TRAP)
- Each of the above tasks must be coded to use an <u>adaptive loop algorithm</u> that checks the accuracy of the calculation after first n sub-intervals, then by increasing the number of sub-intervals by some number (e.g. n+20)), computing the relative change in the solution
- Perform each calculation and report the number of sub-intervals required to achieve a relative error of 1.0E-06 according to

$$\varepsilon = \frac{\left| Integral_{n+5} - Integral_n \right|}{\left| Integral_n \right|}$$

6c. Test your code <u>for each method</u> with the calculation of fractional solid angle when $d = 5$ cm, $R_d = 4$ cm, $R_s = 1.0$ cm, then $\Omega_{fraction} = 0.1073$ and use a convergence of 1.0E-06.

Compute the solid angle and plot your result with d varying from 4 cm to 20 cm, $R_d = 4$ cm, $R_s = 1.0$ cm, and report the number of sub-intervals required for each method.

7. **Waiting Room Dose.** There is a waiting room on the opposite side of a very large wall adjacent to an X-ray Linac treatment room at the local hospital. Compute the flux of photo-neutrons (neutrons generated by high energy photons) $\phi(x)$ into the waiting room using a diffusion theory approximation.

Assume neutrons are emitted from the wall surface via a uniform planar surface source emitting S_o n/cm²/s.

At wall surf. $x = 0$ …

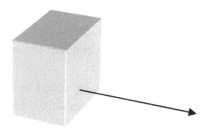

The diffusion equation is $D\dfrac{d^2\phi}{dx^2} - \sigma_a\phi = 0$ for $x \neq 0$

Assuming a 1-dimensional flux approximation and other dimensions of the room relative to the wall (at $x=0$) are <u>infinite</u>, use the following conditions:

    (i) Flux must remain finite as $x\to$ infinity
   (ii) The neutron wall current (coming out of the wall) has a limit as $x\to0$, where:

$$\lim_{x\to0} J(x)\cdot\hat{i} \Rightarrow \lim_{x\to0}\left(-D\frac{d\phi}{dx}\right) = \left(\frac{S_o}{2}\right)$$

  (iii) $D,\ \sigma_a, S_o$ are *constants*

8. **Gamma Analysis.** As an IAEA Safeguards inspector, you collect a gamma ray survey from the surface of a large, heavy container using a Ge semiconductor detector. From the detector's multi-channel readout, you obtain the following spectrum for a 1 minute count, on which you then mark the energies of the associated peaks in keV, based on the detector energy calibration.

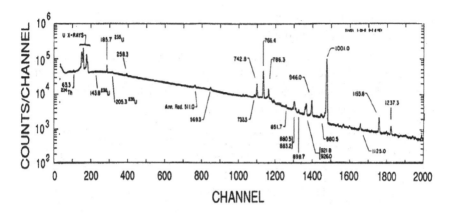

You determine that this is the radiation signature of depleted uranium (DU), which is typically ~0.2% U-235, with the remainder as U-238 (T½ = 4.470E+09 y). Most of these peaks are caused by the aged uranium decay daughter products in equilibrium with U-238, many principally from Pa-234m.

8a. One interpretation of particle flux ϕ is that it is the number of particles moving in any direction through a unit area per unit of time. If we consider the radiation leaving the surface of the container to approximate a point source emanating particles outward without scattering, as we move away from the source the particle flux will drop due to spherical divergence. Assuming the detector used to collect the spectrum is 10% efficient (overall), estimate the total particle flux of 1001.0 keV photons at 1, 5, 10, and 50 cm away from the container. Give your answers in photons/(cm² s).

8b. If the probability per decay of U-238 for the 1001.0 keV photopeak is 6.090E-03, assuming the detector collects 10% of all of the 1001.0 keV photon radiation streaming out of the radioactive sample, estimate the mass present in the container. Assume no self shielding of these gamma rays occurs in the source mass (neglect self shielding).

9. **Mean Values.** Given the transmitted gamma-ray intensity $I(x) = I_o e^{-\mu x}$ in gamma ray transmission experiments, the fraction

transmitted ("transmission") is computed from $T(x) = I(x)/I_o$ considering primary beam attenuation.

9a. Using the figure below, estimate the attenuation constant $\mu \to \mu(E)$ in units of cm^{-1} for gamma rays passing through lead sheets at energies of 200 keV, 400 keV, & 1000 keV:

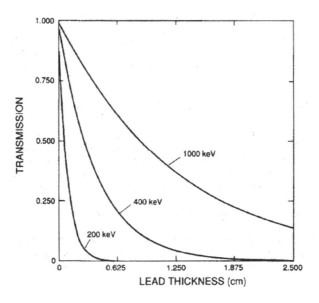

9b. High energy X-ray cargo scanners are designed so that X-rays penetrate varying amounts of lead shielding, for no shielding up to 1 inch thick, so as to penetrate pieces of luggage of various shapes and sizes and be able to render a transmission X-ray image.

Use the above figure with a "by hand" Monte Carlo method to estimate the average transmission value given by $\overline{T} = \dfrac{\int T(x)dx}{\int dx}$ for gamma rays penetrating lead thicknesses spanning from 0 to 1 inch thick. Since the Monte Carlo method must always includes an uncertainty estimate, use <u>two sets</u> of sampling histories (e.g. two sets of "by-hand darts") to estimate an uncertainty in your answer.

10. **Distributions.** The Maxwell-Boltzmann distribution can be used to describe the particle energy distribution of a gas in thermodynamic equilibrium for N_o .particles per unit volume, at an absolute temperature T:

$$N(E)dE = \frac{2\pi N_o}{(\pi kT)^{3/2}} E^{1/2} \exp(-E/kT)dE$$

10a. Show that the most probable energy is $\frac{1}{2}kT$.

10b. Many car dealers use pure nitrogen gas (assume an ideal gas) to inflate car tires. Consider a tire at 50 psia at a temperature of 300K that forms a cylinder of outer radius 76 cm, inner radius 56 cm, and height 25 cm. (The gas constant $R = 8.314472$ Pa.m^3/(mol.K); Boltzmann's constant is 8.617385E-5 eV/K).

Determine the following using a TK-Solver model:

- Volume of nitrogen in the tire
- Number of moles of nitrogen in the tire
- Number of atoms of nitrogen
- Molecular density of molecules if all nitrogen is in the form of N_2
- Plot of $N(E)/N_o$ versus E *(use an adequate "list fill" for sampling points)*
- Most probable energy in eV
- Mean energy computed using the INTEGRAL('function_name, lower_lim, upper_lim) function in TK Solver (uses adaptive Gauss Quadrature); choose an upper limit of integration that is *reasonable* for a numerical estimate of

$$\overline{E} = \frac{\int EN(E)dE}{\int N(E)dE}$$

Set #II

1. **Legendre Polynomials.** Rodrigues' general formula for the Legendre polynomials is:

$$P_n(x) = \frac{1}{2^n n!} \frac{d^n}{dx^n} (x^2 - 1)^n$$

Use this to generate the first four (4) Legendre Polynomials.

Legendre Polynomial	Method	n subintervals	Solution	Relative error
P1	2-Point Gauss, 1 int	1		
P1	2-Point Gauss, n int			
P1	Trapezoidal Rule			
P2	2-Point Gauss, 1 int	1		
P2	2-Point Gauss, n int			
P2	Trapezoidal Rule			
P3	2-Point Gauss, 1 int	1		
P3	2-Point Gauss, n int			
P3	Trapezoidal Rule			
P4	2-Point Gauss, 1 int	1		
P4	2-Point Gauss, n int			
P4	Trapezoidal Rule			

1b. Write a computer application that performs the following using functions:

(i)   *2-point Gaussian quadrature* on a <u>single</u> interval between [0, *a*].

(ii)   *2-point Gaussian quadrature* on *n* <u>intervals</u> between [0, *a*] with a tolerance $\varepsilon = 1.0E - 05$

(iii)   *Trapezoidal Rule quadrature* on *n* <u>intervals</u> between [0, *a*] with a tolerance $\varepsilon = 1.0E - 05$

Then, using the Legendre Polynomials you derived using Rodrigues' formula in part (1a), use the quadrature procedure functions (Gaussian and Trapezoidal rules) verify the orthogonality of the Legendre functions over the interval $[-1,1]$.

Provide the results for the solution of the following integral, where $\phi_i(x) = P_i(x)$ when $i = j$, with $[a,b] = [-1,1]$. Perform this for $i = j = 1$, 2, 3, and 4.

$$\int_a^b \phi_i(x)\phi_j(x)\frac{(2i+1)}{2}dx = \delta_{ij}$$

Be sure to include ALL procedure code, variable, and rule sheets. Be sure to fill in your results in the Table provided.

2. **Applying Newton's Method.** As Chief of Radiation Engineering, you are required to provide assessments on hazards from radiation sources. One of your lab workers requires the use of a uniform spherical Mo-99 source in her laboratory experiment—this isotope yields gamma rays exceeding 140 keV in energy as it decays with a half life of 65.94 hours.

In spherical sources, many gamma rays are attenuated by the mass of the source itself; this is known as the principle of "self-shielding." To determine the exact leakage of gamma rays from a particular geometry, one must ordinarily perform a *radiation transport calculation*. However, several analytical methods are available to bound the amount of actual radiation exiting the source to account for self-shielding. The fraction of gamma rays that emerge (leak) uncollided (not colliding) from a spherical source can be estimated using the following analytic equation:

$$F(X) = \frac{3}{2X}\left[1 - \frac{2}{X^2} + \exp[-X]\left(\frac{2}{X} + \frac{2}{X^2}\right)\right] \quad \text{where } (X = \rho\mu d)$$

$\rho = 10.2$ g/cc for Mo

$\mu = 0.47$ cm^2/g for Mo at ~140 keV, and

d is the diameter of the sphere in cm.

Your Problem: Use Newton's Method coded in your own *TKSolver procedure function* to determine the diameter of the Mo-99 source sphere so that only 5% (*F*=0.05) of the gamma rays contained in the source leak uncollided.

Show all work, and provide a solution that is good to at least 6 significant digits. Report the number of iterations required.

3. **The Deadly Tea.** An MI-5 British Security Service officer enters a bar in London as an undercover infiltrator of a terrorist cell. The barkeep is paid by a double agent to serve a pot of "special" tea to the officer; the tea is later found to be laced with the isotope Polonium-210 (at wt = 209.98286 amu, T½ = 138.376 days).

During the 1 hour meeting in the bar, the officer drinks the tea, which contains 0.5 micrograms of Po-210 per ml, consuming the tea at a rate of 100 ml/hr. At the same time, his body processes the ingested isotope, eliminating it through a well mixed gastrointestinal volume of 1000 ml at a rate of 90 ml/hr through elimination processes.

The net remaining metabolized Polonium-210 that stays in the body emits an energetic alpha particle (5.3 MeV per disintegration) that deposits the full decay energy immediately in local tissue, causing extreme doses.

The activity of Po-210 at any time t is given by:

$$N(t)\lambda = A(t) = A_o \exp(-\lambda t) \quad \text{dis/s}$$

The total number of disintegrations is in the interval of time from $[0,\tau]$ is

$$\text{Total disintegrations} = \int_0^\tau A(t)dt$$

Po-210 distributes itself almost uniformly throughout 10 major organs in the body, including the liver and kidneys, where each organ can be assumed to have an average mass of 1.0 kg. Also, it can be assumed that 40% of all ingested Po-210 remains trapped, long term, in each organ. A novel treatment using diethylene triamine pentaacetic acid (DTPA) chelating agent can be administered in the hospital to remove Po-210 within 72 hours of the initial ingestion.

3a. Identify how Po-210 is produced. Please be specific.

3b. Based on the information given, *neglecting any decay*, compute the *average* amount of Po-210 in micrograms in the MI5 officer over the first hour? Using this average mass as the amount, *neglecting isotope decay over this first hour*, compute the dose to the average organ during the first hour.

3c. Based on the total mass deposited at the end of the first hour when the officer stopped drinking the tea, now accounting for decay, if the lethal dose to each organ is 5 Gy (1 Gy = 1 J/kg), can the officer survive in the next 72 hour period before the hospital chelating treatment takes effect? If not, when does the dose become lethal?

3d. What is the total dose over the first 14 days?

Use any package/library function/programming language you wish to solve this problem.

4. **Peak to Average Power.** A cylindrical nuclear reactor has an axial (z-axis) power profile given by:

$$q'(z) = q'_{max} \cos\left(\frac{\pi z}{H_2}\right) \quad \text{W/cm,}$$

where $-\frac{H_1}{2} \le z \le +\frac{H_1}{2}$.

The two dimensions given are related by $H_1 = 0.95 H_2$, and total power is $Q_{tot} = q'_{avg} \cdot H_1$. Determine the ratio of $\frac{q'_{max}}{q'_{avg}}$ (known as an axial *hot channel factor*).

5. **Radial Heat Transfer.** You are the chief engineer on a Turkmenistanian submarine powered by a pressurized water reactor with a rated full power of 10 MWt. Fuel pins in the reactor are in the form of enriched annular fuel, with an inner and outer cylindrical radius, very *tightly* wrapped with aluminum cladding around the outer radius. Water coolant circulates outside of each fuel pin, constituting a "fuel channel." A typical fuel channel schematic is provided below.

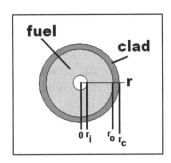

Other data for this reactor are as follows:

– Heat is only generated in the nuclear fuel due to fission
– Heat is transferred **only radially** out of the fuel elements, and there is no temperature gradient in the central gap
– Temperature is a maximum at the fuel inner radius
– No. of Fuel elements (in channels): 50

Al cladding radius: r_c =1.6 cm is at temperature T_c

Fuel outer radius: r_o =1.5 cm is at temperature T_o

Fuel inner radius: r_i =0.1 cm is at temperature T_i

Fuel Length: L = 50 cm

Clad conductivity: k_c =1.65 W/(cm C)

Type "A" Fuel conductivity: k_f = 0.276 W/(cm C)

Type "B" Fuel conductivity: k_f = 0.400 W/(cm C)

Peak/Average hot fuel channel power ("hot channel") factor: 1.8

Type A and B both have a fuel melting temperature of 3000 C; note that we must never design a reactor where the fuel melts!

Steady State heat transfer in the fuel region is governed by:

$$\frac{1}{r}\frac{d}{dr}(r\frac{dT}{dr})+\frac{q_f{'''}}{k_f}=0$$

Where T is temperature, $q_f{'''}$ is volumetric fuel heat generation in W/cm^3

– Heat is conducted across the fuel surface along a unit vector $\hat{r}$ according to Fourier's law:

$$q_f{''}=-k_f\frac{dT}{dr}|_{r\to r_o}$$

where $q_f{''}$ is the heat flux out of the fuel surface in W/cm^2

– Heat exiting the fuel surface is then conducted through the clad along $\hat{r}$ according to:

$$q_c'' = -k_c \frac{dT}{dr}\Big|_{r \to r_c}$$

where q_c'' is the heat flux out of the cladding in W/cm^2

So that the heat transferred <u>must</u> be conserved according to the following **balances**:

$$q_f'''(\pi(r_o^2 - r_i^2)L) = q_f''(2\pi r_o L) = q_c''(2\pi r_c L) = q_c' L$$

Determine the following using a model constructed by you in the package of your choice (TKSolver, Fortran, etc):

5a. For Fuel A, compute q_f''' for an *average fuel channel* and the *"hottest" fuel channel*

5b. For Fuel A, Solve the heat equation in the *hottest channel* (attach all work!) using the notation above to determine the <u>temperature difference</u> ($T_i - T_o$) in degrees C across the *hottest fuel pin*; provide sheet printouts for a TKSolver or equivalent model, or source code.

5c. For Fuel A, Determine q_f'' in the hottest channel (use either heat balance or Fourier's law!)

5d. For Fuel A, determine q_c'' and q_c' in the hottest channel

5e. For Fuel A, if the clad surface temperature T_c is 400 C, determine min and max fuel temperatures in the hottest channel

5f. Again using if $T_c = 400$ C, using Fuel B, determine the maximum fuel temperature.

5g. If the captain orders you to exceed the maximum reactor power to try to escape from an American submarine, how high of a reactor power level can you go to (to avoid fuel melting) if the reactor is loaded with fuel type A? How about if the reactor is loaded with fuel type B?

6. **Photoneutrons.** The neutron spectrum of (γ, n) photoneutrons (from *J. Rad. Analytical Nuc. Chem., Vol 283, pp. 261-265*) has two components shown in the equation below, composed of *the evaporation spectrum* and the *direct emission spectrum*:

$$n(E) = A\frac{E}{T^2} Exp[-E/T] + B\frac{\ln[E_{max}/(E+S)]}{\int_0^{E_{max}-S} \ln[E_{max}/(E+S)]\, dE}$$

"Evaporation neutron" component + "knock-on neutron" component

Where:

-   A and B are two normalization factors to account for the evaporation neutrons, A, and knock-on neutrons, B.
-   T is the nuclear temperature of a particular nucleus
-   E is the energy of neutrons
-   E_{max} is maximum photon energy
-   S is the average neutron binding energy in the nucleus (for targets with multiple isotopes)

For a target anode made of tungsten (W), as used in most all linear accelerators (Linacs),

$$A = 0.8929,\ B = 0.1071,\ T = 0.5\ \text{MeV, and}\ S = 7.34\ \text{MeV}.$$

6a. Verify the average binding energy S of the last neutron in an average Tungsten target nucleus using *TK Solver*. Assume tungsten has the following naturally occurring isotopes (with abundance):

W-180, (.0012), W-182 (.265), W-183 (.143, 1.1E+17 T½), W-184 (0.3068, 3E+17 Y T½), W-186 (0.284)

Hint: This is done by comparing the mass-energy difference of (the final nuclear mass of an isotope having one less neutron plus a neutron mass added) minus (the nuclear mass of the isotope). One could set this up in TK Solver with Lists, output using a table after list solving. Note there are 931.49 MeV/amu

6b. Compare your value $S_{Computed}$ to the literature value ($S_{Literature} = 7.34$ MeV).

$$\text{Use Relative accuracy; } \varepsilon = \frac{\left| S_{computed} - S_{Literature} \right|}{S_{Literature}};$$

6c. Plot the photoneutron distribution spectrum $n(E)$ using *TK Solver* for 8, 16, and 24 MeV photons colliding with W.

7. **Mass and Triple Integration.** The differential volume element in spherical coordinates (r, θ, φ) is given by:

$$dV = r^2 \sin\theta \; d\theta \; d\varphi \; dr$$

The density of a uranium slug is

$$\rho(x, y, z) = \frac{10}{(x^2 + y^2 + z^2)^{1/2}} \text{ g/cm}^3$$

Problem: Convert the following integral, defined in *Cartesian coordinates*, over the volume of the slug (so as to compute the mass of uranium), to an integral in *spherical coordinates*. Then, evaluate that integral.

$$\text{mass} = \int_0^1 \int_{-\sqrt{1-x^2}}^{+\sqrt{1-x^2}} \int_{-\sqrt{1-x^2-z^2}}^{+\sqrt{1-x^2-z^2}} \rho(x, y, z) dy dz dx$$

Set #III

1. **Quadrature.** Consider the following 2nd degree real polynomial:

$$f(x) = a_0 + a_1 x + a_2 x^2$$

1a. Integrate the polynomial on the real interval from [A,B] to yield I, the exact integral. Show ALL work.

1b. Use the Midpoint Rule for numerical quadrature on 2 equal subintervals (denoted as M_2) on the real domain [A,B]. Show ALL work.

1c. Determine $\varepsilon_{M2} = (M_2 - I)$ (the error in the 2-subinterval Midpoint quadrature obtained by subtracting your solution in 1a from your solution in 1b). Show ALL work.

1d. Given that in a similar manner as used in 1b and 1c, we computed error for the 2-subinterval Trapezoidal rule

$$\varepsilon_{T2} = (T_2 - I) = \left[-\frac{1}{24}(A - B)^3 a_2 \right],$$

determine the value of the constants c_1, c_2 that will render a combined estimate of the integral $S_2 = c_1 M_2 + c_2 T_2$ as accurately as possible. Show ALL work.

2. **Reduction of Order.** Obtain a solution t_0 to $y'' + 2\sqrt{2}y' + 2y = 0$.

Then, use *Reduction of Order* to obtain a second solution. Show ALL Work.

3. **Frobenius' Method.** Find the general solution of

$$4x^2 y'' - xy' + (1-x) y = 0$$

(You *must* use the method of Frobenius—show all work, listing either the recurrence relation or at least the first *three* terms of any series solutions). Show ALL work.

4. **Plutonium Production.** Pu production comes from neutron irradiation of U-238 in a reactor. The major fissile isotopes of plutonium are Pu-239 and Pu-240. Pu-239 is fissile, and although Pu-240 decays by spontaneous fission, it mainly captures in neutron irradiation. While Pu-241 is also fissile, this is produced in small amounts from capture in Pu-240; for simplification, we will not track Pu-241 production here. Consider a mass of initially pure depleted uranium (pure U-238) introduced in a reactor. Assume the following:

– The initial mass of the U-238 is 1000 g, at a density of 19.8 g/cc

– The steady neutron flux throughout the mass is 1.0E+14 n/ cm²/s

– Nuclide Data:

uranium-238,	$T_{1/2} = 4.46\,E9\,y$	$\sigma_{a28} = 2.4\,E\text{-}24\,cm^2$ (assuming all reactions are capture)
neptunium-239,	$T_{1/2} = 2.35\,d$	$\sigma_{a39} = 53.6\,E\text{-}24\,cm^2$
plutonium-239,	$T_{1/2} = 2.41\,E4\,y$	$\sigma_{a49} = 976\,E\text{-}24\,cm^2$
		$\alpha_{49} = 0.362$ (this is the capture (n,γ) to fission ratio)
plutonium-240,	$T_{1/2} = 6537\,y$	$\sigma_{a40} = 280\,E\text{-}24\,cm^2$

The nuclides are related by the following differential equations, where each N refers to atoms of material:

$$\frac{d N^{28}}{d t} = -N^{28}\,\sigma_{a28}\,\phi$$

$$\frac{d N^{39}}{d t} = +N^{28}\,\sigma_{a28}\,\phi \;-\; N^{39}(\lambda_{39} + \sigma_{a39}\,\phi)$$

$$\frac{d N^{49}}{d t} = +N^{39}\,\lambda_{39} \;-\; N^{49}(\lambda_{49} + \sigma_{a49}\,\phi)$$

$$\frac{d N^{40}}{d t} = +N^{39}\,\sigma_{a39}\,\phi + \left(\frac{\alpha_{49}}{1+\alpha_{49}}\right)(N^{49}\,\sigma_{a49}\,\phi) - N^{40}(\lambda_{40} + \sigma_{a40}\,\phi)$$

Problem: Solve the set of differential equations *by hand*, Then, model the solutions in TKSolver and plot the weight fraction of each nuclide using *TKSolver* for total neutron irradiation spanning from 0 to 300 days.

5. Solve the Euler-Cauchy Differential Equation

$$x^2 y'' + 4xy' + 2y = x \cos(\ln(x))$$

6. **Reflected Critical Mass.** You are working in a joint counter-terrorism unit with the Homeland Security Department in Miami. During a routine inspection, airport security forces have discovered a solid sphere of Pu-239 of radius R = 5.5 cm and coated in very thin plastic, disguised as a child's toy ball. The FBI wants to ship the ball back to a Department of Energy (DOE) lab for evaluation. Since there is a Coast Guard Cutter docked at a nearby port, they plan to send the Pu ball packed in a crate on board the ship. An FBI man secures a very large crate full of small foam packing peanuts (in the center of which he plans to carefully place the Pu sphere for shipping).

The Steady-State one energy Neutron Diffusion Equation with Neutron Multiplication in the Pu ball (region 1) is:

$$D_1 \nabla^2 \phi_1 - \sigma_{a1} \phi_1 + v\sigma_f \phi_1 / k = 0$$

The Steady-State one energy Neutron Diffusion Equation in the packing peanuts (region 2) is:

$$D_2 \nabla^2 \phi_2 - \sigma_{a2} \phi_2 = 0$$

where in both regions

$$\nabla^2 \phi = \frac{1}{r^2} \frac{\partial}{\partial r} \left(r^2 \frac{\partial \phi}{\partial r} \right)$$

assuming spherical symmetry.

To transform the equation into something that is solvable, a change of variables necessary to transform this into a linear equation with constant coefficients is The following nuclear data may be useful:

For the Plutonium Ball (we'll neglect the plastic coating):

$$D_1 = 1.040 \text{ cm} \qquad \sigma_{a1} = 0.0949 \text{ cm}^{-1} \qquad v\sigma_{f1} = 0.255 \text{ cm}^{-1}$$

For the Foam Packing Peanuts Used (values are approximate):

Dow Type "A" Packing Peanuts $D_2 \approx 2.5 \text{ cm}$ $\sigma_{a2} \approx 10^{-5} \text{ cm}^{-1}$ $v\sigma_{f2} = 0 \text{ cm}^{-1}$
Dow Type "B" Packing Peanuts $D_2 \approx 3.0 \text{ cm}$ $\sigma_{a2} \approx 10^{-5} \text{ cm}^{-1}$ $v\sigma_{f2} = 0 \text{ cm}^{-1}$

Background: The steady state neutron diffusion (Helmholtz) equation (above) represents a balance of leakage, absorption, and production from fission reactions in the sphere using the fundamental-mode eigenvalue. When we apply the BARE SPHERE extrapolated boundary condition that: $\phi(R+2D_1)=0$ at the radius of the sphere (R) plus the nuclear "extrapolation" dimension ($2D_1$) (for a diffusion physics correction), we are essentially assuming that no neutrons are reflected back into the sphere. In fact, because atmospheric air reflects back few neutrons (~zero) that escape the sphere surface, this is a good assumption; therefore, we need not consider solving the diffusion equation considering the air immediately surrounding the sphere.

However, if we replace the material surrounding the Pu sphere with a hydrocarbon (such as foam packing peanuts), we could get some very nice reflection of neutrons back into the sphere, and we will have to account for the diffusion of neutrons in the region of the foam peanuts outside of the sphere.

The multiplication factor (k) of the 5.5 cm radius unreflected ("bare") sphere in open air is $k = 0.9322$. This means that the steady state loss rate of neutrons by absorption and leakage from the sphere is greater than the rate neutrons are produced inside the sphere from fission. If there is enough material present to ramp up production from fission reactions to make k = 1.0, then a steady state balance of production from fission and loss due to absorption and leakage can occur—known as a "criticality"... Notice how the "k" in the steady state diffusion equation "adjusts" the production term to satisfy the overall equation. When k exactly equals 1.0, criticality is achieved, and the mass causing this condition is known as a "critical mass"—not a good thing to be standing next to, since the radiation coming directly in the local vicinity of a critical mass is lethal.

6a. Solve for the critical size (when k = 1.0) of a "bare sphere" (use the fundamental mode of the solution).

6b. Determine the critical size (k = 1.0) for a sphere reflected by an infinite thickness of packing peanuts (assumed since the crate is large). Based on your answer, should the FBI man pack the 5.5 cm radius sphere in the crate packed with which packing peanuts, if any, Dow Type "A" or Dow Type "B" (e.g. which ones allow for more safety with regard to criticality?)

To solve this part, the following conditions apply:

(i)   The plutonium ball (region 1, spanning r = [0,R]) is a multiplying medium and the flux in the ball is finite at the origin
(ii) The packing peanut region (region 2, spanning r=[R,∞)) constitutes a non-multiplying medium, is assumed to be infinitely thick, and the neutron flux is finite as the radius r increases; this is made up of packing peanut types Dow "A" or Dow "B".
(iii) The material interface region (where r = R) between the Pu and the foam packing peanuts must have BOTH continuity of flux and continuity of current.

Once these conditions are applied, use TKSolver or other language to root solve for the minimum positive radius of plutonium R for $k = 1$. Then, plot the solution for the flux using TKSolver (you may assume the arbitrary ball flux coefficient is 1.0).

6c. Repeat part b to determine the critical size ($k = 1.0$) for a sphere reflected by an outer (total) radius of 15 cm of packing peanuts. Should the FBI man pack the 5.5 cm radius sphere in this smaller crate?

6d. Demonstrate that $k = 0.9322$ for the 5.5 cm radius "bare sphere"

Set #IV

1. The inverse of a matrix A can be found by initially setting the matrix A with the identity matrix in the form $[\ A\ |\ \ I\ ]$, where I is the identity matrix (I is a matrix with all zeros except '1's are placed into each diagonal location). Then, using standard row operations, the position of the identity matrix is shifted, leaving the inverse of the matrix in the position originally occupied by the Identity matrix, where $[\ I\ |\ A^{-1}\ ]$. The inverse matrix can then be used to solve the system $A\,x = b$ by performing the multiplication step $x = A^{-1}\,b$.

Problem: Write a computer code (using your choice of FORTRAN, C, C++, or TKSolver (for TK, see "How To Import and Export Data" in Help) to determine the inverse of an n x n matrix A using Row Operations.

Specifically, this should be a code that YOU write to solve the system A $x = b$ by first finding the matrix inverse, and solving the system by performing the multiplication step $x = A^{-1}\ b$. *Solution steps should include:*

(i) Read in an n x n matrix A from a text file.
(ii) Read in one or more column vectors b from the text file.
(iii) Perform *row operations* on A to shift the matrix from $[\ A\ |\ I\ ]$ to $[\ I\ |\ A^{-1}\ ]$
(iv) Compute the solution vector(s) x based on one or more sets of b vectors
(v) Print your solution vector(s) in a clear way to an output file called "result.out"

1a. Include your code with your solution vectors for the following:

$$A = \begin{pmatrix} 8 & 3 & 7 & 1 \\ 1 & 7 & -1 & 2 \\ 2 & 5 & 10 & 5 \\ 2 & 4 & 1 & 6 \end{pmatrix}$$

$$b1 = \begin{pmatrix} 7 \\ 12 \\ 4 \\ 16 \end{pmatrix} \qquad b2 = \begin{pmatrix} 64 \\ 18 \\ 136 \\ 160 \end{pmatrix}$$

1b. Include your code with your solution vectors for solving the following:

$$A = \begin{pmatrix} 21 & 31 & 71 & 24 \\ 10 & 40 & -10 & 20 \\ 22 & 25 & 21 & 30 \\ 28 & 42 & 43 & 60 \end{pmatrix}$$

$$b1 = \begin{pmatrix} 41 \\ 38 \\ 54 \\ 16 \end{pmatrix} \qquad b2 = \begin{pmatrix} 64 \\ 29 \\ 100 \\ 60 \end{pmatrix}$$

2. **Cryptography and Matrices.** Cryptography (Coded messages) has a basis in matrix algebra. Consider a "Message" matrix M, a crypto "Key" matrix K, and an "Encrypted message" matrix E which can result from the matrix multiplication of M and K:

$$M \cdot K = E$$

2a. **Problem:** If the "Key" matrix K is given as indicated below, use the row operations method to determine the inverse matrix K^{-1}.

$$K = \begin{pmatrix} -1 & 0 & 1 \\ 2 & 3 & 4 \\ 2 & 4 & 5 \end{pmatrix}$$

Once an encrypted message E is delivered to the recipient, it can be decoded by multiplying by the inverse key matrix K^{-1} applied on the right half of each side of the equations, yielding:

$$M \cdot K \cdot K^{-1} = M \cdot I = M = E \cdot K^{-1}$$

With this knowledge, and supplied with the following inputs as assumed "known values":

- a "Key" matrix K <u>and</u> its inverse K^{-1}
- a code "Translation" matrix T (to translate numbers into corresponding letters and spaces) given by:

$T =$

$$\begin{pmatrix} 12 & 17 & 8 & 21 & 27 & 19 & 3 & 25 & 20 & 11 & 5 & 26 & 7 & 23 & 1 & 13 & 10 & 24 & 6 & 2 & 16 & 4 & 18 & 15 & 22 & 9 & 14 \\ a & b & c & d & e & f & g & h & i & j & k & l & m & n & o & p & q & r & s & t & u & v & w & x & y & z & \end{pmatrix}$$

Write a computer code to *decode <u>and translate</u>* the messages E #1 and E #2 using matrix multiplication methods to accomplish $(E \cdot K^{-1})$, assuming that K^{-1} is already a known provided from above, to determine M #1 and M #2. What do the messages say?

2b. Encoded Message #1 (left) and #2 (right)

$$\begin{pmatrix} 98 & 173 & 225 \\ 38 & 98 & 138 \\ 33 & 94 & 133 \\ 76 & 162 & 221 \\ 26 & 62 & 84 \\ 72 & 147 & 201 \\ -6 & 22 & 49 \\ 33 & 98 & 149 \end{pmatrix} \qquad \begin{pmatrix} 43 & 99 & 131 \\ 47 & 134 & 198 \\ 52 & 128 & 178 \\ 45 & 132 & 195 \\ 14 & 43 & 71 \\ 84 & 177 & 246 \end{pmatrix}$$

2c. **Problem:** If the "Key" matrix K is given as indicated below, use the row operations method to determine the inverse matrix K^{-1}.

$$K = \begin{pmatrix} 4 & 9 & 2 \\ 3 & 5 & 7 \\ 8 & 1 & 6 \end{pmatrix}$$

For Part 2b, with this knowledge, and supplied with the following inputs as assumed "known values":

- a "Key" matrix K <u>and</u> its inverse K^{-1}
- a code "Translation" matrix T (to translate numbers into corresponding letters or a space) given by:

$T =$

$$\begin{pmatrix} 12 & 17 & 8 & 21 & 27 & 19 & 3 & 25 & 20 & 11 & 6 & 2 & 16 & 4 & 18 & 15 & 22 & 9 & 14 & 5 & 26 & 7 & 23 & 1 & 13 & 10 & 24 \\ a & b & c & d & e & f & g & h & i & j & k & l & m & n & o & p & q & r & s & t & u & v & w & x & y & z & \end{pmatrix}$$

Decode the following: Encoded Messages #3 (left) and #4 (right)

$$\begin{pmatrix} 264 & 226 & 260 \\ 287 & 194 & 329 \\ 228 & 149 & 178 \\ 381 & 358 & 371 \\ 242 & 214 & 234 \\ 280 & 180 & 320 \end{pmatrix} \qquad \begin{pmatrix} 133 & 184 & 223 \\ 124 & 227 & 174 \\ 251 & 134 & 245 \\ 285 & 356 & 319 \\ 188 & 177 & 250 \\ 307 & 314 & 309 \\ 142 & 220 & 178 \end{pmatrix}$$

3. **Gram–Schmidt Procedure.**

Given that :

$$f(x) = \sum_{n=0}^{\infty} a_n P_n(x) = a_0 P_0 + a_1 P_1 + a_2 P_2 + \ldots \qquad \text{on } [A, B]$$

With moments

$$a_n = \left[c_n \int_A^B f(x) P_n dx \right] \qquad \text{and} \qquad c_n \equiv \left(\int_A^B (P_n(x))^2 dx \right)^{-1}$$

Using the interval $[0, \Delta x]$ with the Gram–Schmidt procedures/ Weierstrass's Theorem, it can be shown that:

$$P_0(x) = 1 \qquad\qquad\qquad c_0 = \frac{1}{\Delta x}$$

$$P_1(x) = x - \frac{\Delta x}{2} \qquad\qquad c_1 = \frac{12}{\Delta x^3}$$

$$P_2(x) = x^2 - x\Delta x + \frac{\Delta x^2}{6} \qquad c_2 = \frac{180}{\Delta x^5}$$

3a. Use the procedures discussed in the text with the above polynomials orthogonal on the interval $[0, \Delta x]$ to find zeroth, first, and second moments for the function $f(x) = \exp[3x/(2\Delta x)] - 1$ with $\Delta x = 2$.

3b. Compare your result to the second order Taylor's series expanded from $x = 1$. Check values compared to the original function at $x = 0$ and $x = 2$.

Show all steps and work.

Set #V

1. The Oven Door Design.

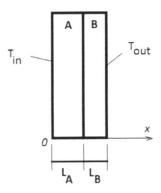

An oven door must be designed from two types of high temperature composites (Type A: $k_A = 0.15$ W/(cm K), Type B: $k_B = 0.08$ W/(cm K)) so that it is thick enough to insulate the user during steady state oven operation, where the outside surface temperature T_{out} must remain at 293K.

The geometry to consider is given in the figure on the right, where y- and z-directions insulated, so heat only flows through the area normal to the x-axis. The heat flux incident on the inner door surface is q'' = 20 W/cm².

Determine the dimensions L_A and L_B of the door if the design calls for $L_A = 3/2\ L_B$ and the oven temperature must be able to safely reach a maximum of 900 K at surface T_{in}.

Plot the temperature profile for the final solution.

2. 2-D Heat Conduction. Solving the Steady State Heat Equation on a flat plate:

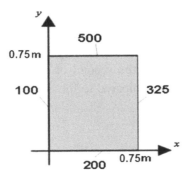

A very thin (z-direction insulated) rectangular copper sheet defined by $x \in [0, L]$, where $L = 0.75$m length along the x-dimension, and $y \in [0, W]$, where $W = 0.75$m width along the y-dimension, has four of its lateral sides held at fixed temperatures as shown (in degrees C). There are no sources or sinks on the plate, so that the governing equation for steady state heat flow through the plate is given by LaPlace's equation, $\nabla^2 T = 0$.

2a. Solve for the temperature distribution T(x, y) of the plate.

2b. Give the temperature of the plate at (x, y) = (0.5m, 0.5m) that considers as many series solution terms as needed to achieve within +/− 0.003 degrees C of accuracy.

3. **Numerical Heat Conduction.** Numerically Solving the Steady State Heat Equation on a flat plate:

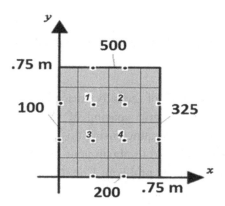

The thin plate from problem 1 will now require a numerical solution using 4 internal temperature nodes. Again, the governing equation for steady state heat flow through the plate is given by LaPlace's equation, $\nabla^2 T = 0$.

3a. Using 4 internally spaced temperature nodes (separated uniformly by 0.25 m), write the (4x4) matrix containing a linear system of equations (in the form $A\vec{x} = \vec{b}$) needed to solve for the temperature at each node (T_1, T_2, T_3, T_4) using the differencing formulations (based on derivations in class):

$$\frac{\partial^2 \phi_o}{\partial x^2} \approx \frac{\phi_W - 2\phi_o + \phi_E}{\Delta x^2}$$

$$\frac{\partial^2 \phi_o}{\partial y^2} \approx \frac{\phi_N - 2\phi_o + \phi_S}{\Delta y^2}$$

3b. Solve the system of equations in part 2a for the 4x4 matrix using matrix algebra methods to invert the matrix and solve for a the temperature at each node. If you wrote your matrix inversion program from the last assignment, you may use that.

3c. Compare the temperature value at $(x, y) = (0.50$ m, 0.50 m$)$ to the value of $T(x, y)$ you found by analytically solving this PDE in question #1.

3d. *Increase* the number of equally spaced gridpoints *along each direction* to yield a mesh grid spacing of $\Delta x = \Delta y = 0.125$ m, and compare the resulting temperatures using this higher resolution grid with the lower resolution grid used to obtain your results in 2a, and compared to the analytic solution in problem 1 at $(x, y) = (0.50$ m, 0.50 m$)$. Comment on solution accuracy.

4. **Derivation of Differencing Equations.** Using the method of Taylor's series expansions analogous to that presented in class, derive the finite difference formulation for node ϕ_0 using ϕ_E and ϕ_{EE} that could be used from a left handed endpoint when solving Helmholtz's equation in one-dimensional Cartesian slab (x) geometry.

Be sure to state the <u>finite difference equation</u> *and* <u>the associated truncation error</u> of the formulation, indicating the order of the truncation error (1st order, 2nd order, etc).

Note that the nodes (shown below-left) are separated by a constant grid spacing Δx, and that Helmholtz's equation is given by:

$$\nabla^2 \phi + B^2 \phi = 0$$

Left boundary Node Schematic along the x-axis

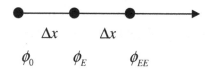

$$\Delta x \qquad \Delta x$$

$$\phi_0 \qquad \phi_E \qquad \phi_{EE}$$

5. Time Dependent Conduction in Metal Rods.

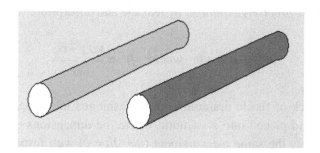

Two metal rods, one made of Gold (Au), one made of iron (Fe), are of length $L = 0.5$ m. These are stored for a long period of time in a freezer at 2 degrees C. The rods are insulated along their length, so that heat transfer can only occur through the ends of the rod. Suddenly, the rods are removed from the freezer and, the both ends are held at a fixed temperature of 200 degrees C. The governing equation for steady state heat flow through the plate is given by the time dependent heat equation.

$$\frac{\partial^2 T}{\partial x^2} = \frac{1}{\alpha}\frac{\partial T}{\partial t}$$

5a. Using separation of variables, solve for the *analytic formulation* of the temperature distribution $T(x, t)$ of a rod.

5b. Look up the properties of Au and Fe to determine a constant value of thermal diffusivity α for each, using units of m^2/s. Use a computer program (FORTRAN, C, etc) or computer package (such as *TKSolver* or *Mathematica*) of your choice to answer this question, provided that YOU (or you *and* your team member) supply the algorithm/logic to perform the computation. How long (in seconds) must one wait for the center of the rod to reach 150 degrees C for each rod, and which one heats up the slowest?

6. **Criticality Using Diffusion.** Given: A critical, steady state neutron multiplying system can be described by the steady neutron diffusion equation. Note the system is exactly critical when fission neutron production is *precisely balanced* by leakage and absorption, with:

$$\nabla^2\phi + B^2\phi = 0 \quad \text{where} \quad B^2 = \frac{\nu\sigma_f - \sigma_a}{D}$$

A solid block of fissile uranium is to be fashioned into a parallelepiped geometry and placed into a vacuum, where the dimensions of each side bear precisely the same measurement ($a = 2b = c$) with respect to the x, y, and z axes.

For the uranium block: $D = 1.1$ cm

$$\nu\sigma_f = 0.1687 \ 1/cm$$

$$\sigma_a = 0.08 \ 1/cm$$

Using Separation of Variables, solve the PDE to determine the critical size *based on the fundamental mode* for a steady state critical assembly. In doing so, use a Cartesian coordinate system (x, y, z) with the origin located at the center of the cube, and assume the flux is zero at an extrapolated distance of $2D$ added to each physical boundary, as routinely performed for a "vacuum" boundary treatment.

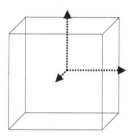

7. The Flux Trap.

7a. Use the *steady state* 1-speed diffusion equation to analytically <u>solve for the critical size, mass (outer radius R2 and height H) and peak flux</u> for a 1 MW cylindrical reactor with a water filled central flux trap located between radius $r = 0$ and $R1$. Plot the flux as a function of r from the center of the reactor (use IF/THEN commands as needed to set appropriate region equations in the plot). The reactor is depicted on the right:

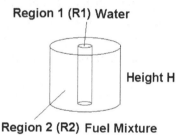

Region 1 (R1) Water

Height H

Region 2 (R2) Fuel Mixture

Boundary conditions:

Vacuum on top ($+H/2$),

bottom ($-H/2$), and **outer radius** ($R2$) of the reactor. The fundamental mode (n = 1) for an exactly critical steady state reactor may be assumed.

Other design constraints include:

- Extrapolated dimensions $\tilde{H} = \dfrac{3}{2}\tilde{R}_2$; Fundamental mode.

- Use the *fuel region* extrapolation distance for *entire* top and bottom surfaces, and $R_1 = 0.65R_2$.
- Cross sections and other parameters in regions (1) and (2): Macroscopic temperature corrected cross sections in *1/cm*, diffusion constants in *cm*, densities in *g/cc*:

Water	**Fuel**
$\nu\Sigma_{f1} = 0$	$\nu\Sigma_{f2} = 0.09435$
$\Sigma_{a1} = 0.022$	$\Sigma_{a2} = 0.035$
$D_1 = 0.243$	$D_2 = 2.115$
$\rho_1 = 0.95$	$\rho_2 = 11.0$
$\Sigma_{f1} = 0$	$\Sigma_{f2} = 0.03883$

7b. Switch the fuel and water regions in Problem 1 to make a 1 MW *water reflected* finite cylinder reactor, and re-solve the necessary equations for the critical size, mass, and peak flux as in problem 1. Plot the flux as a function of r from the center of the reactor.

Compare/contrast your solution with problem 1.

Bibliography

Arfken, G., and Weber, H., *Mathematical Methods for Physicists*, 4th ed, Academic Press, New York, 1995.

Boas, M., *Mathematical Methods in the Physical Sciences*, 3rd ed, Wiley, New York, 1983.

Byron, F., and Fuller, R., *Mathematics of Classical and Quantum Physics*, Vol I and II, Dover Publications, New York, 1970.

Bronson, R., *Modern Introductory Differential Equations*, McGraw Hill Co, New York, 1973.

Currie, L. A. *Analytical Chemistry*, 40 (3), 586, 1968.

Duderstadt, J., and Hamilton, L., *Nuclear Reactor Analysis*, Wiley, New York, 1976.

Farlow, S., *Partial Differential Equations for Scientists and Engineers*, Dover Publications, 1993.

Francois, *Nuclear Instruments and Methods*, Vol. 117: 1974, pp. 153-156.

Glasstone, S., Sesonske, A., *Nuclear Reactor Engineering*, 3rd ed, Krieger Publishing Co, Malabar, Florida, 1991.

Kreyszig, E., *Advanced Engineering Mathematics*, 7th ed, Wiley, New York, 1993.

Lamarsh, J., Introduction to Nuclear Engineering, 2nd ed, Addison-Wesley Publishers, Reading, Mass., 1983.

Lewis and Miller, *Computational Methods of Neutron Transport*, ANS Publications, 1993.

Moore, G. E., "Cramming more components onto integrated circuits," *Electronics Magazine*, p. 4, 1965.

Nelson, D. (Ed), *Penguin Dictionary of Mathematics*, 3rd ed, Penguin Books, New York, 2003.

Nelson, D. (Ed), *Penguin Dictionary of Statistics*, Penguin Books, New York, 2004.

Rhodes, R., *The Making of the Atomic Bomb*, Simon and Schuster Publishers, 1986.

Schmidt, F., "ME-540 Course Notes—Numerical Solution of PDEs," Penn State University, 1994.

Swokowski, E., *Calculus with Analytic Geometry*, 2nd ed, PWS Publishing, 1979.

Zill, D. G., and Cullen, M. R., *Advanced Engineering Mathematics*, 2nd ed, Jones and Bartlett, Boston, Mass., 2000.

Index

Printed in the United States
By Bookmasters